Ecological Vignettes

Ecological Vignettes

Ecological Approaches to Dealing with Human Predicaments

Eugene Odum

University of Georgia
Institute of Ecology
Athens, Georgia, USA

with cartoons by Sidney Harris and Tom Hammond

harwood academic publishers

Australia Canada China France Germany India
Japan Luxembourg Malaysia The Netherlands
Russia Singapore Switzerland Thailand

Amsteldijk 166
1st Floor
1079 LH Amsterdam
The Netherlands

Illustrations by Julie Latham.

British Library Cataloguing in Publication Data

Odum, Eugene P. (Eugene Pleasants)
  Ecological vignettes : ecological approaches to dealing
  with human predicaments
  1.Environmental protection   2.Human ecology
  I.Title
  333.7'2

ISBN   90-5702-521-3

To my late wife *Martha Ann*, landscape painter,
and our late son *William Eugene*,
aquatic ecologist and ardent fisherman

Contents

You can't see the whole picture when you're standing inside the frame. This person sees only the healthy part of the world environment.

It was only after astronauts took pictures from outer space that we saw the whole earth. This one is from the moon. Courtesy of NASA.

Preface

Just about everything that concerns us as humans on this earth — growth, competition and cooperation, health, quality of life, survival, use of energy, space and resources, relationships with other forms of life, and so on — has parallels in nature. Therefore, it stands to reason that we can learn a lot from ecology, the science of the environment, that can help us understand and deal with our human predicaments and dilemmas. After all, organisms and organized communities of organisms (ecosystems) have been surviving and prospering on earth through the thick and thin of cosmic and geological upheavals much longer than human society. The idea of the "wisdom of nature" is not new, of course, but I believe that it is only with the recent development of a holistic or ecosystem ecology that we can really apply what we are learning from environmental studies to increase chances that the human race will not only survive but continue to flourish, despite the litany of threatening problems and hazards.

In this book I present some basic ecological principles in the form of vignettes — memorable, provocative statements designed to attract your attention at least long enough for you to read the explanations. We hope that presenting ecological concepts in this way will help avoid overpopulation, over-consumption, damage to life-support systems and other suicidal behavior. In presenting the "wisdom of nature" in this manner, we are not projecting nature as some sort of all-knowing "guru," but simply inquiring how life has not only survived but thrived over millions of years through good times and bad. We recognize that not all human predicaments have parallels in nature; some are unique to human society. We consider some of these, again in the form of vignettes, in chapter 7.

Until recently environmental literacy was not considered a major part of our cultural heritage; very little environmental education was taught in schools beyond, perhaps, a bit of nature study or geography. Our slowness to act to halt unsustainable population growth and to preserve the quality of our environment may be in part due to the fact that most people in the current decision-making generation are environmentally illiterate. It is encouraging that environmental education is beginning, finally, to receive attention.

As shown in the cartoon on page xi, it is time that we step out of the frame and consider the whole picture of the earth and its physical and biological conditions as related to human affairs. Also, as suggested by the cartoon on page xv, just accumulating money will not ensure a quality future for humankind. Let us take our noses out of our wallets and take a look at the big picture of humans and that all-important life-supporting environment.

If this book helps even in a very small way to improve environmental literacy at all levels — kindergarten through senior citizenry — then it will have served the purpose intended.

I am indebted to the staff and students of the Institute of Ecology at the University of Georgia for ideas and encouragement over the past 40 years.

Acknowledgments

Thanks to Sidney Harris for permission to reproduce several cartoons that appeared in his book *There Goes the Neighborhood: Cartoons on the Environment*, University of Georgia Press, Athens, and to Tom Hammond, University of Georgia, Art Department, for cartoons especially prepared for this book.

TAKE YOUR NOSE OUT OF YOUR WALLET AND LOOK AT THE BIG PICTURE.

CHAPTER 1

WHAT WE LEARN FROM ECOLOGY ABOUT GROWTH

Vignette 1.1 ————————————————————————

TO GROW OR NOT TO GROW IS NOT THE QUESTION. THE QUESTION IS WHEN TO STOP GETTING BIGGER AND START GETTING BETTER.

Explanation. How often have you heard the statement "grow or die"? Is this a true statement of fact, or should we add "but not under all circumstances"? To the businessperson, continuous economic growth is an article of faith; without such growth, business will fail. But when growth in body size of the individual stops at adolescence, one does not die. In fact, to many of us the best part of life begins! Are there different kinds of growth, such a quantity versus quality growth? Are there times and places where growth is appropriate, and are there times and places where growth is detrimental (like cancer)? To get some reasoned answers to these questions let us consider the basics of growth forms.

THE CARRYING CAPACITY CONCEPT AND THE S-CURVE

The basic pattern for the growth of almost anything is a **sigmoid** curve, as shown in Figure 1.1, in which we plot increase in size (weight, numbers, or other measures of size) on the vertical axis and time on the horizontal axis. With this pattern, the rate of growth is slowed or reduced more and more as size approaches some sort of upper limit or when some kind of plateau is reached. Such a pattern is often known as an "S-curve" for growth. For an individual, the limit to growth is determined by the genetic makeup, which halts growth when full adult size is reached. For populations and ecosystems, the limit is what ecologists call the carrying capacity, that is, the size that can be sustained at a given time and place. Leveling off at the population or

1

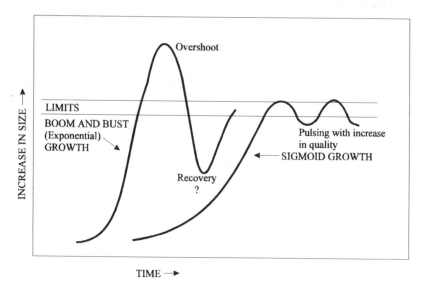

Figure 1.1

system level is not automatic or certain, as it is with the individual, but depends on when diminishing return of scale (negative feedback) starts to reduce the growth rate. Usually, the carrying capacity plateau is a pulsing one where size varies up and down around a plateau level, depending on fluctuations in the environment or other external forces. If for any reason the limits are raised, then growth in size may begin to plateau again at a new fluctuating carrying capacity level. Sigmoid growth forms, either self-limited or moderated by interactions with other species (competitors, predators, or parasites, for example), are the most common patterns that we find in nature.

However, as also shown in Figure 1.1, there is another quite different pattern of growth that is not uncommon. It involves a more or less uncontrolled or exponential rate of increase, with doubling and redoubling at short time intervals, which results in a boom-and-bust pattern (Cartoon 1.1), because the momentum becomes so great that size overshoots the carrying capacity level and a rapid decrease or "downsizing" occurs. Some natural populations (e.g., Arctic lemmings) exhibit this kind of growth; numbers increase very rapidly until the population overshoots some resource limit or becomes victim to predation or disease. Or maybe a favorable season or other conditions end. Then the population "crashes," with the death of large numbers of individuals, perhaps to repeat the cycle at some later time. Of

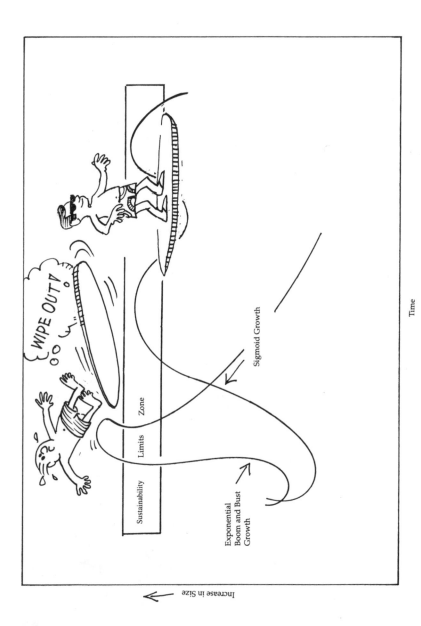

Cartoon 1.1 Riding the waves or wipeout in surfing is analogous to sigmoid vs. boom-and-bust growth models.

course, with this growth pattern there is always a risk of extinction if the "bust" is too deep.

Unfortunately, we are seeing this "boom-and-bust" pattern more and more in human affairs. While it is logical to assume that economic growth, for example, cannot keep increasing without disastrous "busts" in a planet that is itself not growing, there remains a widespread belief that humans are more or less immune to limits because human ingenuity and technology can overcome environmental or resource limitations. There is no real evidence that such is the case, but there are the cornucopia ("horn of plenty" metaphor) technologists among us who are optimistic that hydrogen economy, landless agriculture, wasteless industry, and other new technologies will enable a very large human population (ten billion or more) to coexist with enough natural environment to provide the necessary life support. We will consider some of these possibilities in Chapter Seven, especially as related to the "technological paradox." But first we need to consider other aspects of growth and the key role that energy plays in the human–environment interaction.

HOW SOCIAL INSECTS DEAL WITH GROWTH

Perhaps we can learn something from the social insects: the ants, termites, and social wasps. Like humans, they tend to form large colonies, which superficially resembles cities. Also like humans, these insects have been extremely successful in the evolutionary survival race, since they comprise from 50 to 80 percent of the total biomass (weight) of all insects now living on the planet. Their success is due in large measure to cooperative division of labor within the colony and efficient use of resources. Although their rigid caste system — i.e., specialized workers, reproductive individuals, nurses, soldiers — is not something we would want to emulate, their ability to regulate colony size so as not to exceed environmental limitations is something that might merit our attention. Professor E.O. Wilson, an world authority on the social insects, speaks of colony regulation as "programmed demography" in that, as a colony grows to a large size, birth and death rates are altered so that population size levels off, thereby avoiding "boom and bust" or the syndrome of too rapid growth followed by decline.

Population control in the harvester ant (so-called because they collect and store seeds in underground galleries) is an especially good example. The colonies grow in sigmoid fashion, leveling off in density after about six years. As the number of ants returning to the colony without seeds (signifying that the local food supply is getting scarce) increases, the rate of "antenna contacts" (the way ants communicate with each other) rises, and fewer eggs and larvae are produced (Gordon, 1995). Unfortunately, when it comes to

"YOU KNOW AS WELL AS I DO THAT A MAN IS MEASURED BY THE AMOUNT OF POLLUTION HE CREATES."

Cartoon 1.2

humans, interaction between people in crowded cities seems to increase violence and pollution without a decrease in birth rates!

FUTURE HUMAN POPULATION GROWTH

The future pattern of human population growth will play a major role in the future quality of society and the environment. Most demographers (population scientists) project that human population growth will eventually be sigmoid, as shown in Figure 1.2. Even if every woman now alive bore two children who survived into adulthood, the so-called replacement rate, the human population, now five billion, would continue to grow until well into

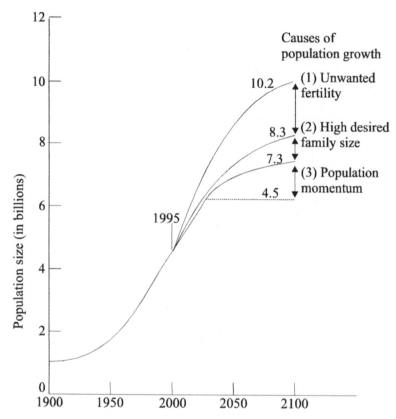

Figure 1.2 Bongaart's projection for human population growth (*Science* **263**:771–776, 1994)

the next century and perhaps level off at a minimum of seven billion. This is "population momentum," the third factor in Figure 1.2, which results from the fact that such a large part of the population of the underdeveloped world is just coming to reproductive age, so that there will be a lot of births even if women on average had only two children. The two other causes of growth shown in the table are "unwanted fertility," babies born who are not wanted or cannot be cared for (teenage pregnancies, for example), and "high desired family size," where children are perceived to be needed for child labor and to take care of parents in their old age. Adding the effects of these three factors projects a leveling off at about ten billion by 2100.

Society cannot do anything about population momentum, but a concerted effort could be made to reduce growth factors 1 and 2 if enough people and political leaders were convinced that such an effort would result in a better world for all. We can be optimistic about this because at the 1994 United Nations conference on population held in Cairo (a city with a very large population of poor people) it was revealed that more and more women in underdeveloped countries are seeking family-planning services in order to have two rather than six children. Furthermore, all nations voted for a plan to commit billions of dollars to the cause of curbing population growth, especially by promoting education and career opportunities for women.

THE OPTIMUM IS LESS THAN THE MAXIMUM

Safe or optimal carrying capacity, as a level below the maximum or saturation level, is a concept that comes out of the ecological study of animal populations that may be relevant to our human situation. It has often been observed that animals such as quail or muskrats maintain population numbers that are well below that which might be supported by food supply and other vital resources. In such cases, individuals are less vulnerable to predators, disease, or weather that temporarily reduces food supply or habitat cover. What is known as **territorial behavior** or the **territorial imperative**, in which individuals and families do not tolerate close neighbors, is another mechanism that we may observe in nature that keeps a population below saturation level.

As a general principle, we can say that in terms of the well-being of the individual, the optimum density is less than the maximum. Or, to put it more bluntly, the world can support more warm bodies, like cows in a feedlot, than it can support quality human beings!

THE QUESTION OF WHEN

Returning to the original question of **when**, we can conclude that there are times and places when growth in size is necessary for survival, and there are situations in which further growth in size is deleterious (i.e., cancerous). Then it is time to stop getting bigger and start getting better. More on this in Vignette 1.3.

When individuals, towns, businesses, or systems in general are small or young, it is generally "grow or die," but when things get large, complex, or mature it may be "grow and die"! For more on the growth dilemma, see Essay 1.

Vignette 1.2

AFFLUENCE REDUCES THE POPULATION SIZE THAT CAN BE SUSTAINED ON A GIVEN RESOURCE BASE.

Explanation. "You cannot have your cake and eat it too" is an old expression that comes back to haunt us as we face a world that is increasingly divided between the rich and the poor. Remember that during the French Revolution Marie Antoinette got her head cut off some time after she was reported as suggesting "let them eat cake" when informed that the poor did not have enough bread. The actual number of humans or other organisms that can be supported on a given resource base for the long term depends not only on environmental carrying capacity (size of the sustainable resource base, etc.) but on **lifestyle**. We observe in the real world of nature that far fewer meat eaters (predators — lions and tigers, for example) than plant eaters (herbivores — antelopes and rabbits, for example) can be sustained on a given area of land. As an old Chinese saying goes, "Only one tiger to a hill." Parallels in human affairs immediately come to mind. Since the per-capita consumption of fossil fuel energy in an affluent country may be 40 times that in a poor country, something like 40 times more people with a poor country lifestyle can be supported on a given level of available fossil fuel. In other words, an affluent lifestyle, like living high on the food chain, greatly reduces carrying capacity density. One widely read author (Garrett Hardin) speaks of this limit function as the "cultural carrying capacity." Eaters of expensive cake will be fewer than eaters of cheap bread. For a case study on "optimum population" at a U.S. lifestyle, see Essay 2.

In summary, the carrying capacity concept is two-dimensional, involving the number of individuals and the intensity of per-capita demands on the environment. These attributes track in a reciprocal manner: as the intensity of per-capita use **goes up**, the number of individuals that can be supported by a given resource base **goes down**. For more on the concept of carrying capacity, see Essay 3.

How to deal with this dilemma is one of the great questions of our age. Do we try to raise the affluence of everyone in the whole world to that of the affluent countries (currently the home of less than a third of the world's population, and projected to be even less in the future), which is probably impossible and certainly impractical? Or would the quality of human life worldwide be better served if the wealthy nations powered down some, so that the poor nations can power up without endangering the necessary life-supporting goods and services (air, water, food, etc.) provided by the natural and agricultural environment? A leveling of the playing field would seem to be achievable without really reducing the standard of living of the rich nations. We can get some clues from nature as to how this might be done, since there are natural ecosystems, such as coral reefs, or forests growing on poor soil, that not only survive but prosper under conditions of scarce resources. How this is achieved is detailed in Essay 4.

Vignette 1.3

QUALITY IN CONTROL OF QUANTITY IS THE GREAT LESSON IN BIOLOGICAL EVOLUTION (LEWIS MUMFORD). IT IS ALSO A GREAT LESSON FOR HUMANITY.

Explanation. As suggested in Vignette 1.1, there is life, perhaps a better life, after growth in size levels off. For want of a better term, we can call this phase "qualitative growth" or "quality growth," which we define as getting better rather then bigger (see Cartoon 1.3). We know from personal experience that quality can follow quantity, since when growth in body size stops at adolescence we devote most of our lives to becoming better, not bigger, human beings. Likewise, when growth in size of a forest levels off, the quality of wood increases but total timber volume does not (which is one reason why timber companies are so eager to engage in boom-and-bust logging of the old-growth stands on our public lands).

I believe we can say without too much contradiction that there are limits to quantitative growth in a finite world, but as far as we can see there need not be any limits to qualitative growth. Quality can always be improved, especially after quantity has had its day.

To top off our discussion, we can say that growth patterns in nature are more or less preordained by genetics at the individual level and by natural selection at the community and ecosystem levels, but that these natural "regulators" are much less effective when it comes to humans and human affairs. We can be optimistic in that we do have choices. We can continue to choose economic and governmental policies that promote more or less unrestricted growth in human populations and the accumulation and con-

Cartoon 1.3

sumption of material wealth that will likely produce huge boom-and-bust oscillations, widen the gap between rich and poor, and damage, perhaps beyond repair, the earth's life-support systems. Alternatively, we can choose policies that promote quality rather than quantity of life; in other words, promote **better** rather than **bigger**. The second choice will require major

Cartoon 1.4 When campus carrying capacity is exceeded, then quality of education declines.

changes in the way we think, behave, and do business. Such changes are now being widely discussed in books and magazines but as yet are not seriously considered as political issues. It is human nature to be conservative and resist changes that might be painful. Most people continue to cling to the idea that bigger is always better (or necessary to maintain a high standard of material wealth), even when there is increasing evidence to the contrary. Again, it is human nature to wait until things get dangerously bad before making changes for the better. We just don't want to wait until it is too late!

There is one human institution, the university, in which administrators recognize that bigger may not be better (as suggested by Cartoon 1.4). It is here that real efforts are being made to make the transition from bigger to better by plateauing total enrollment and increasing the quality of education. Raising admission standards, increasing the quality of teaching and research, raising promotion standards for faculty, diverting less-inspired students to community colleges, and increasing the proportion of graduate and professional students are some of the strategies for achieving quality over quantity. More about this in Essay 5.

CHAPTER 2

WHAT WE LEARN FROM ECOLOGY ABOUT ENERGY

Vignette 2.1

ENERGY, THE COMMON DENOMINATOR THAT CANNOT BE REUSED.

Explanation. In the 1992 presidential campaign, candidate Bill Clinton kept saying to his staff, "It's the economy, stupid," to remind them that the state of the U.S. economy was the main issue at that time. In the future it may well be "It's energy, stupid!" If you were asked to pick out a single common denominator for life on earth, that is, something absolutely essential and involved in every action, large or small, the answer would have to be energy. We use the word "energy" for the ability or capacity to do work, with work defined in the broadest sense of to do or perform something. While you are working at your job or hobby or relaxing, or even sleeping, your body is performing thousands of vital functions that require energy of a specific amount and kind. The fantastic variety of organisms and functions involved in keeping our air, water, soil, and food life-support systems operating requires a great deal of energy.

Early in life we become aware that the primary energy source for green plants is light and the indirect solar energies (rain, wind) required for photosynthesis, and that the primary energy source for humans, animals, and all the other non-green organisms is the food made by the green plants. In addition, societies, especially industrial ones, require large amounts of very concentrated energy in the form of fuels to operate all the human-made machines and infrastructures.

Because energy is vital to everything, there is no excuse for ignorance about the principles of energy transformations. We should start teaching **energetics** in the first grade, or even earlier. It is especially important that we

13

know about and understand the two laws of thermodynamics, and also the relation between quantity and quality in energy transformations, which will be discussed in the next vignette.

THE ENERGY LAWS

The first law of thermodynamics states that energy may be transformed from one form (such as light) to another form (such as food) but is never created or destroyed. The second law states that all transformations are accompanied by a degradation of some of the energy from a concentrated form into a dispersed form such as heat, so that no transformation can be anywhere near 100 percent efficient. The degraded energy, or "entropy," is essentially a "waste" that has to be removed from the system to prevent it from becoming "disorderly." Thus, although energy is neither created nor destroyed during transformations from one form to another, some of it is degraded to a dispersed form (heat, for example) which is no longer available, as shown in the sun-leaf transfer diagram in Figure 2.1.

The more important thing to remember about all this is that energy cannot be reused, in contrast to materials such as water, minerals, or money, which can be recycled or used over and over again with little or no loss of utility (see Cartoon 2.1). The food you ate for breakfast this morning is no longer available to you once it is transformed into maintenance, growth, and repair of your body tissues. You must go to the store and buy more for tomorrow. Organisms, ecosystems, and human-made mechanical systems (automobiles, for example) alike are open systems that require a continuous flow of energy through them and that must continuously "pump out" the disorder (or, in more formal terms, "dissipate the entropy").

In Chapter One we spoke of the "great dilemma" of when to grow bigger and when not to. The great challenge in regard to energy is not only how we will deal with the inherent and increasing disorder that accompanies increasing energy use, but also how we humans will manage when we have used up the nonrenewable fossil fuel energy sources (coal, oil, natural gas) on which modern industrialized civilizations depend? Will we have to resort to some form of atomic energy, with all its disorderly aspects, OR can we make do, as does the natural world, with solar energy, an abundant renewable source? We very likely will need to do some of both.

ENERGETICS

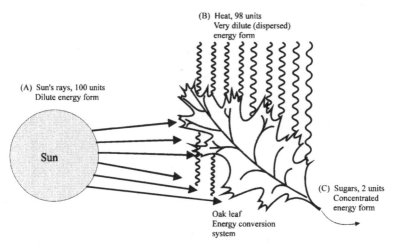

Figure 2.1 The two laws of thermodynamics. The first law is illustrated by the conversion of sun energy (A) to food (sugar, C) by photosynthesis (A = B + C). The second law dictates that (C) is always less than (A) because of heat dissipation (B) during conversion

The problem with solar energy is that it is a very low-quality (dispersed) source that has to be greatly concentrated to provide fuel that is equal to fossil fuel. Sunlight alone won't run your car unless it is greatly concentrated. There are several ways this can be done with known technology. Sunlight can be directly transformed to electricity by photovoltaic cells or by concentrating sunlight with mirrors to run steam turbines. Electricity can also be generated by ocean thermal energy conversion (OTEC) systems, which use the temperature differential between the warm surface and cold deep water in tropical oceans, an indirect use of solar energy.

The electricity generated by any of these methods can then be used to separate hydrogen fuel out of the water. A major advantage of a hydrogen economy is that using hydrogen as a fuel does not release carbon dioxide (CO_2) into the air as does the burning of fossil fuels. Increases in atmospheric CO_2 resulting from human activities is a major factor in current trends toward global warming and other undesirable climate changes. The catch, as is so

Cartoon 2.1   Energy cannot be recycled.

often the case, is that fuel produced by these concentration technologies will likely be more expensive than fossil fuel that has already been concentrated by nature.

Making special efforts to reduce current wasteful uses of fossil fuel will have the dual benefits of reducing air and water pollution right now and giving us more time to work out alternatives for the future. So remember, "It's energy, stupid!" For more on energetics as the common denominator, see Essays 6 and 7.

Vignette 2.2

IN A CHAIN OR SEQUENCE OF ENERGY TRANSFERS, THE QUANTITY OF AVAILABLE ENERGY DECLINES WITH EACH STEP, BUT THE QUALITY MAY BE GREATLY ENHANCED (another "bad news/good news story").

Explanation. The decline in quantity and the increasing disorder potential inherent in energy transfers (as reviewed in Vignette 2.1) can be a negative aspect of thermodynamics; however, there is a bright side since the transformed part is often more concentrated and therefore has a higher "quality" in terms of work potential. Green plants, for example, can convert not much more than 5 percent of light energy to food energy, but food is of a much higher work quality since it will support plant and animal life while unconverted light will not.

Accordingly, in a chain or sequence of transformations such as a food chain (sun → grass → cow → human) or a fossil fuel chain (prehistoric plant → peat → coal → electricity), quantity declines with each transfer but quality increases, as shown in Figure 2.2. In the human food chain, meat (the animal level) is more concentrated (and more expensive) than vegetables (the plant level), which explains why fewer people can be supported on a meat diet as compared to a vegetable one. In the fuel energy chain, it took many calories of prehistoric plant life to make one calorie of coal (concentrated by fossilization), and then several calories of coal are required to make a calorie of electricity, a higher-quality energy source for humans. Accordingly, there are

(A) Pictorial Diagram food chain

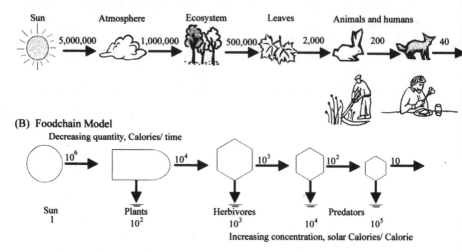

(B) Foodchain Model

(C) Electric Energy Chain

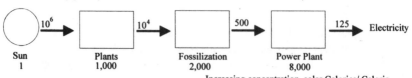

(D) Spatial hierarchy model (dispersed to concentrated)

Figure 2.2 Increasing energy concentration (quality) accompanies decreasing quantity in food chains and in electric energy generation; involving five orders of magnitude (orders of 10). Figures are in kcal/m^2. After H.T. Odum, 1983.

both benefits and costs in energy transformations, which brings us back to the quantity–quality dilemma (it is difficult or impossible to have both at the same time), as discussed in Chapter One.

Vignette 2.3

THE RISING TIDE OF MAINTENANCE COSTS, THE ULTIMATE LIMITATION.

Explanation. As natural ecosystems develop by the process of ecological succession (going from bare ground to a forest, for example), we observe that more and more of the available energy is required to "pump out the disorder" as the ecosystem increases in size, complexity, and diversity. When a point is reached where all of the available energy is required for maintenance, there can be no more net growth in size or function without the risk of collapse due to inadequate maintenance.

In describing or modeling the energetics of this natural ecological development, the total energy available from photosynthesis is called gross primary production, or P, while the maintenance energy is called community respiration, or R. Early in succession (youthful or pioneer stages), as opportunistic species of plants and animal colonize and reproduce rapidly, P exceeds R and the total biomass (living weight) increases until R is equal to or exceeds P (Figure 2.3). In ecology the stage where P = R is often designated the **mature stage** or the **climax**. In general, as size doubles, more than double the maintenance energy is required, so that maintenance costs can build up rapidly.

There are parallels in cities, in that more than double the energy per capita is required to maintain the increased infrastructure when a city doubles in size. This is why taxes increase rapidly as cities become very large. This is not to say that we should not have large cities, since they do have cultural advantages, such as museums and art centers, business centers, professional sports, and so on, not possible in small towns. But there is a price to pay in that large cities are very expensive to maintain, and prone to disorder. High local taxes and/or other subsidies from the outside (the federal government, for example) become a necessity. A formula for the relationship between size and maintenance costs is explained in the fourth principle of Essay 1.

A brief comment on taxation, always a political issue, might be in order at this point. To tax or not to tax is not the question, because there have to be some taxes to pay for the collective values of civilization. The larger and more complex civilizations become, the more money is required to "maintain order" or "domestic tranquillity," the term used in the preamble of the U.S. Constitution. The question should be: how do we use our wealth most efficiently to maintain or improve quality of life? For example, to combat

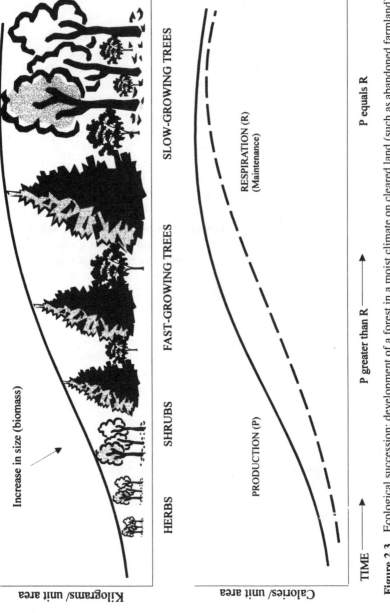

Figure 2.3 Ecological succession: development of a forest in a moist climate on cleared land (such as abandoned farmland) showing changes in structure and metabolism

crime, do we spend large amounts of our tax monies building prisons to warehouse criminals, or do we spend it on social and economic services designed to see that the children of the poor do not become criminals?

Civilizations, past and present, have difficulty dealing with the increasing tendency for government bureaucracy to proliferate with size, making it difficult to use tax monies efficiently for the public benefit. Geographer Karl Butzer states this problem as follows: "Civilizations become unstable and break down when the high cost of maintenance results in a bureaucracy that makes excessive demands on the productive sector." As we noted in Chapter One, nature's solution to the "bureaucracy problem" has involved the slowing down of growth in size as safe carrying capacity is approached, rather than waiting until it has been exceeded.

Eric Hoffer, a self-styled seaman philosopher who has traveled the earth, wrote that he could judge the quality of a country by how well the facilities in the ports of call were maintained. Good maintenance of infrastructure indicates good quality of human life. Or, conversely, deterioration of infrastructure and services such as sanitation, repair, crime prevention, and health care is a warning sign that quality of life in that country is not what it should be.

CHAPTER 3

WHAT WE LEARN FROM
ECOLOGY ABOUT ORGANIZATION

Vignette 3.1 ——————————————————————

THE BASIC ORGANIZATIONAL PLAN FOR BOTH
HUMANS AND NATURE INVOLVES HIERARCHIES
CONSISTING OF SUCCESSIVE LEVELS OF ORGANI-
ZATION.

Explanation. A hierarchy is defined as arrangement into a series
of graded compartments. Picture, for instance, Chinese boxes,
with a box inside a box inside a box, and so on. We can illustrate
by comparing geographical, military, and ecological levels of
organization hierarchy, as shown in Table 3.1. The ecological and
geographical hierarchies are said to be "nested" in that each level
is made up of a group of next-level-below units. Thus, states are
made up of groups of counties and landscapes groups of ecosystems. In contrast, military organization, in common with many
other human organized hierarchies, is "non-nested"; sergeants, for
example, are not composed of groups of privates. These non-
nested arrangements tend to be more rigid than the nested ones.
Society might do better if our human organizations were less rigid
and more flexible and interactive as in nature's.

In the ecological arrangement, as shown in Table 3.1, the ecosystem is the
lowest level that has all the parts necessary for a functional system, namely,
all the organisms and physical factors that interact in a given area. For this
reason, the ecosystem level provides a logical point to start an analysis of the
environmental situation or problem.

Ecosystems are very much open systems, with energy, organisms, and
materials continually coming in and going out. Accordingly, "inputs" and
"outputs" are integral parts of a functional system. Higher levels in the life
system's hierarchy, such as landscapes, biomes, and the biosphere, are levels
involving interaction of an increasing number and variety of ecosystems.

Ecological and physiological	Geographical and political	Military
Ecosphere or Biosphere	World	General
Biogeographic Region	Continent	Colonel
Biome (biotic region)	Nation	Major
Landscape	Region	Captain
Ecosystem	State (or Province)	Lieutenant
Biotic Community	County	Sergeant
Population (species)	Town (or Township)	Private
Organism	Human Population (ethic, etc.)	
Organs	Individual	
Tissues		
Cells		
Genes		

Table 3.1   Examples of levels of organization hierarchies

Because scientific research tends to be "reductionist" (i.e., focusing on parts rather than wholes), we know more about genes, cells, organs, and organisms than we know about populations, ecosystems, landscapes, and the ecosphere. As will be noted in the next vignette, the higher levels are more than just sums of their parts, so we cannot understand and deal with these higher levels by just studying the parts; we also need to study these higher levels as wholes, and this is what the science of ecology is all about.

Each level in the ecological hierarchy influences what goes on at adjacent levels. Processes at lower levels are often constrained in some way by those at higher levels. For example, a parasite and its host frequently engage in an oscillating "arms race" as each seeks to get the better of the other. But operating within the larger systems over the long term, parasites and hosts, as well as predators and prey, tend to adjust to one another to achieve some sort of balanced coexistence. In other words, large systems tend to be less changeable than the smaller component ones. Humans invite increasing trouble with parasites, pests, and diseases by promoting rapid population and/or economic growth that not only exceeds the environmental carrying capacity of the landscape and biome, but also negates the self-organizing balancing processes that are emergent properties of large systems.

ECOSYSTEM MANAGEMENT

It is becoming evident that natural capital (natural resources) and human-made capital (market resources) have to be managed as a whole since they are interdependent. Accordingly, the concept of "ecosystem management" that considers both resources in making decisions is being projected as a wave of the future (see Essay 8). At the present time, watersheds (i.e., all the land that drains into a stream, river, or lake) are proving to be good units for study and management, as is discussed in Essay 9.

THE GAIA HYPOTHESIS

The theory that the biosphere has functioned as an organized system since the very beginning of life is known as the Gaia Hypothesis, "Gaia" being the Greek goddess "Mother Earth." According to this theory, living organisms over the ages have not just adapted to the physical environment but have actively modified the environment and "co-evolved" with it to produce the current high-oxygen atmosphere and the complex and diversified ecosystems of today. The earth's first atmosphere did not contain any oxygen but lots of carbon dioxide like today's atmosphere on nearby Venus and Mars. On earth, the first green bacteria that appeared more than two billion years ago began putting oxygen into the atmosphere, and early marine bacteria began removing carbon dioxide to form limestone. These changes in the atmosphere made it possible for "oxygen breathers" (aerobes) like ourselves to take dominion.

The Gaia Hypothesis, first outlined by physicist James Lovelock and microbiologist Lynn Margulis (see Lovelock's 1979 book, *Gaia: A New Look at Life on Earth*) remains controversial in scientific circles. However, we can pose the following question. Are humans going to be the first life forms to render asunder what nonhuman life forms (or God, if you prefer) have created?

Vignette 3.2

THE WHOLE IS GREATER THAN THE SUM OF ITS PARTS.

Explanation. When a person exhibits an overly narrow viewpoint on some subject, we are often tempted to remark, "He or she can't see the forest for the trees." A basic theory underlying the whole-is-greater-than-the-part proposition is the **emergent property principle**. As parts are organized, either naturally or by human design, to produce larger functional wholes, new properties emerge that are not present in any of the parts. For example, when two gases — hydrogen and oxygen — are combined in a certain molecular configuration, water is formed, a liquid with new properties quite different from those of the two original components.

An example from the living world is coral. When certain algae and coelenterate animals evolve together to form a coral, an efficient plant–animal nutrient cycling whole is produced that enables the coral reef to maintain a high productivity in nutrient-poor seawaters, as is detailed in Essay 4. Since neither component can do very well by itself, the ability of the partnership to prosper in an infertile environment is an emergent property. Nature is replete with such successful partnerships, as we will detail in Chapter Five.

While just about everyone recognizes that our body is more than a collection of organs, society as a whole has been slow to recognize that humans are a part of, not apart from, the earth's life-supporting physical and biological systems. As we continue to take dominion over the earth, our future well-being will depend more and more on how successful we are in becoming stewards of the earth, that is, taking care of and repairing our life-supporting environmental systems. For more on stewardship, see Essay 25.

The bottom line to all this is that parts of the biosphere, such as vegetation and humans, do not prosper, and may not even survive, if the whole biosphere is in bad shape. The part depends on the whole just as the whole depends on the part. The concept that the natural areas are necessary components of our total environment is outlined in Essay 10.

Essay 11, based on an address presented at a conference of insurance executives, summarizes much of what we have covered in Chapters Two and Three.

CHAPTER 4

WHAT WE LEARN FROM ECOLOGY ABOUT CHANGE

Vignette 4.1

THERE ARE CHECKS AND BALANCES BUT NO EQUILIBRIA IN NATURE.

Explanation. In past years, we heard a lot about the "balance of nature" and how such and such might "upset" this balance. We now understand that, while there are very important balances in nature, such as the balance between atmospheric oxygen and carbon dioxide that has persisted for aeons, there is no such thing as a central control device (like a thermostat) that keeps nature as a whole in equilibrium.

As we noted in the preceding chapter, many relationships are very dynamic, that is, changeable or pulsing back and forth as in the case of parasites and hosts, each striving to outwit the other. For any given space-time, it is often difficult to predict what Mother Nature is up to, so to speak, because any given event or process may or may not interact with other processes or events, many of which seem to happen at random, that is, by chance. Yet overall, on the large scale of the oceans, the land masses, and the atmosphere, there are checks and balances that maintain a pulsing status quo over long periods of time, and many patterns of growth and development are predictable.

It all comes down to a question of **cybernetics**, the name we give to the science of controls. Bodily functions such as body temperature are controlled by a center in the brain set to maintain body temperature within very narrow limits. Likewise, control of temperature in your home or car is accomplished by a set-point device that we call a thermostat. In human affairs, we often

27

employ regulatory devices such as the Federal Reserve Board, which attempts to control inflation by setting interest rates. It is these kinds of set-point controls that are covered by the standard dictionary definition of cybernetics as "the science dealing with comparative study of human control systems, as the brain and nervous system, and complex electronic systems."

As we have already hinted, nature operates under an entirely different kind of cybernetics. Control is effected by feedback within the system, specifically by the action–reaction of positive and negative feedback, rather than by set-point mechanisms. Feedback occurs when some part of the output energy or a signal is "fed back" into the system to either accelerate (positive feedback) or decrease (negative feedback) some function. In a general way we outlined how the action–reaction system works in our description of the sigmoid growth pattern (Figure 1.1). In Vignette 2.3 we discussed how growth in size stops when maintenance energy is equal to or exceeds production energy. Since there is no set-point in time or place when such balances might occur, they can vary widely and are therefore hard to predict, but sooner or later all processes have their point of diminishing returns.

Scientists like to coin Greek words for basic concepts, and we do have two Greek words for the two kinds of control we have been discussing. **Homeostasis** (maintaining a steady state) is a good word for control in the organism and in the levels below in the hierarchy, while **homorhesis** (maintaining the flow) is a good term for non–set-point cybernetics at levels above the organism.

So what do we learn from nature about control and development? We learn that feedback control (as we observe it in nature), not set-point control (by dictators, for example), must be the modus operandi for humanity. We have discovered that this kind of control does work, since our forefathers in the United States, after some trial and error, designed a government in which feedback (i.e., action and counteraction) between the President, Congress, and the courts is designed to keep a reasonable balance between extremes (i.e., keep political right–left swings from getting out of hand). The problem is that we have not yet learned how to include environmental and quality considerations in societal cybernetics. In other words, we have not learned how to recognize when diminishing returns of scale set in.

Vignette 4.2

YOUTH-TO-MATURITY CHANGES IN NATURAL SYSTEMS HAVE PARALLELS IN HUMAN AFFAIRS.

Explanation. In Vignette 2.3 we discussed how energy use changes in ecological succession, which is the natural process of ecosystem development and repair. In Vignette 4.1 we noted that there is a lot of design as well as uncertainty in nature. Now we wish to look at ecosystem development as a predictable youth-to-maturity design with great relevance to human affairs.

As previously noted, development of natural communities on sites that have abundant unused resources begins when opportunistic species colonize, reproduce rapidly, and build up an organic structure. The size and composition of this structure depend on climate and the physical resources at the site, as well as on resources that may be imported to the site from adjacent communities. This period of succession is often known as the pioneer stage. As community size (biomass, or total amount of life) increases, crowding and pressure on resources results in a slowdown in size development. There is a shift into a mature stage during which pioneer species are replaced by those more adapted to crowded conditions. In ecology, the mature stage is often known as the "climax," but this may not be the most appropriate word since it suggests a stationary condition. Quite the contrary, the climax stage may be very dynamic because of the interplay of forces and counterforces. Its just that size remains roughly within the limits imposed by energy, resources, and space (i.e., the carrying capacity).

The youth-to-maturity development that we see in nature has interesting parallels in the development of the individual that we all experience, and in the development of societies as they are faced with crowding and scarce resources. In the individual, the transition, which we call "adolescence," is controlled genetically, so we go from childhood to adulthood at a certain age whether we like it or not. In natural and human communities, the transition is not so automatic, but results, as we have already explained, from negative feedback, with no set point as to when it will occur. In all cases, many processes — such as high birth or growth rates, which are appropriate during youthful stages — become inappropriate at mature stages, so that transitions of this kind require an about-face on many functions. In the development of natural communities, almost no pioneer species are able to make the transition and are, therefore, replaced by other species. If the human species is to weather the transition, there has to be a change in lifestyle. In other words, the pioneer "mind-set" has to shift to the mature "mind-set." One thing we can agree on is that it is human nature to try to prolong youth for as long as possible! For more on youth–maturity concepts, see Essay 12.

In summary, if we humans are to survive the demographic youth-to-maturity transition we must, among other things, learn how to control growth (as discussed in Chapter One) and deal with the increasing costs in energy and resources required to maintain order when population density and human-made infrastructures (such a cities) become very large and complex. Also, we need to recognize the importance of the kind of cooperation and mutual aid that we observe in successful natural communities that manage to prosper when resources become scarce. More about this in the next chapter.

As far as what we learn from nature is concerned, the only alternative to such a worldwide youth-to-maturity demographic transition is to alternate the youth and mature states. Such a strategy would involve abandoning or downsizing communities that have exceeded sustainability to make way for another cycle of youthful growth. This, of course, is what happens when a mature forest is destroyed by a storm, opening the way for the return of opportunistic "youthful" growth-oriented species. In human society such a pulsing situation might be accomplished without an increase in mortality by promoting a period of negative growth (natural deaths exceeding births). There are activist groups in the United States that are now promoting the negative population growth idea (see Cartoon 4.1).

"WHEN I STARTED WORKING HERE, THERE
WERE ONLY THREE PEOPLE IN THIS OFFICE."

Cartoon 4.1

CHAPTER 5

WHAT WE LEARN FROM ECOLOGY ABOUT BEHAVIOR

Vignette 5.1 ——————————————————————

WHEN THINGS GET TOUGH, IT PAYS TO COOPERATE.

Explanation. Although survival of the fittest in nature does involve a lot of confrontation, competition, predation, and other kinds of violent behavior, there is also a lot of peaceful cooperation, not only between individuals, but between species as well, for mutual benefit. Ecologists use the term **mutualism** when the partnership between unrelated species is so strong that neither partner can survive without the other. Mutualism very often develops between two organisms that have something that the other needs, so that forming a partnership is mutually beneficial. For example, there are numerous cases where animals and non-green plants, such as fungi and many bacteria, form mutualistic partnerships with green plants that provide food in return for the mineral nutrients or protection that the non-green organisms can provide.

An example of a fungal–algal partnership in the lichen, a plant that thrives in the toughest environments, including rock outcrops and Arctic tundras. The tough fungi form a protective net over the algae, which provide food for the fungus. The fact that different kinds of lichens have evolved independently in many different families of fungi indicates how important this kind of cooperation is to the survival of the non-green organisms over geological time.

Another very important mutualism is the root fungus–tree partnership. The root fungus, or mycorrhizae, rather than being a parasite or decomposer of organic matter as are many fungi, is a helpmate that assists the roots in extracting minerals from very poor soil in return for food transported from the green leaves down to the roots. Pine seedlings inoculated in the tree nursery with a large population of mycorrhizae can grow in places where nearly all the topsoil has eroded away. Foresters use this combination to rehabilitate badly damaged lands.

As was noted in Chapter One and discussed in Essay 4, a coral reef is an example of a natural ecosystem where green plant–animal mutualism is so efficient in use and recycling of nutrients that the community prospers in nutrient-poor tropical waters. Since the coral animal has algae growing in its body, we can say that it grows its vegetables in its "belly" during the day, when the sun is shining, and fishes for a little meat by extending its tentacles out into the water at night! In all likelihood, we will need to develop such an arrangement in future cities, where vegetables are grown on rooftop green-houses, with meat and other non-green items mostly imported from outside.

Legumes and some other plants are able to "fix" nitrogen, that is, convert gaseous nitrogen (which cannot be used by higher plants) to nitrates (which can be) only because they form a partnership with nitrogen-fixing bacteria. These plants form root nodules that house and feed the symbiotic bacteria, which in return fertilize the soil. Another very important mutualism involves microorganisms that can digest cellulose and animals that cannot. The cow is a familiar example. Well-chewed grass is deposited in the cow's rumen, where microorganisms break down the plant's fibers into carbohydrates and fats that the cow can use.

We could compile a very large book just by describing all the partnerships that are found in nature, most of which directly or indirectly benefit humans.

Many people think of Charles Darwin and his theory of natural selection in terms of the poet Tennyson's "red tooth and claw" metaphor. Actually, Darwin was quite aware of nonviolent natural selection and wrote about it in his two best-known books, *The Origin of Species* and the *Descent of Man*. But it was a Russian named Peter Kropotkin who first emphasized the importance of cooperation in his book *Mutual Aid*, published in 1895, some years after Darwin. Because Kropotkin was labeled a communist, his ac-counts of the parallels between cooperation in nature and in human affairs were not taken seriously until recently. In a nutshell, his contribution to theory was to suggest two basic kinds of natural selection: (1) organism vs. organism, which often leads to confrontation and competition, and (2) organisms vs. environment, which leads to cooperation and mutualism. In other words, to survive an organism does not confront or compete with its environment but must adapt or modify its environment in a cooperative manner. Developing a mutually beneficial partnership with other organisms helps a great many species to survive exacting environments. And so, ultimately, must humans.

The most important relationship between non-green humans and green plants is agriculture. Until recently this relationship has been more exploitative than mutualistic, that is, we have focused on increasing harvest with little regard to the decline in soil quality and increase in water pollution that occurs when only yield is considered. To counter this obvious non-beneficial relationship, new agricultural procedures, variously called "reduced input agriculture," "conservation tillage," and "residue management," are coming into greater use. These practices strive to reduce the amount of fertilizer, pesticide, and plowing and to increase the amount organic residue retained on the land in order to reduce soil erosion and increase water-holding capacity, thereby making the human–plant relationship more sustainable over time. For more on the input management concept, see Essay 13.

Ants and acacia trees provide an interesting story that is relevant to human dilemmas as to when to fight and when to cooperate. African acacia trees house ants in special cavities in the branches and feed them with nutritious excreted sap. In return, the ants protect the tree from would-be insect herbivores. If the ants are experimentally removed (by poisoning with an insecticide, for example), the tree is quickly attacked and often killed by defoliating insects. In some parts of the tropics of the New World, acacias do not form partnerships with ants, but instead defend themselves with anti-herbivore chemicals. So here we have two options for surviving in a contentious world: form cooperative coalitions or develop expensive defensive weapons.

We all hope and pray that the recent dramatic shift in the relationship between the superpowers is the beginning of a shift from confrontation to cooperation similar to the one we see in nature when resources and energy become scarce. For several decades, the United States and the Soviet Union increased their production of weapons in the name of defense. As the cost of this confrontation began to take an appreciable percentage of the wealth of each county, and as the defense initiative became a powerful political means to win elections or maintain dictators, consumer, social, and environmental needs were neglected. As the pressure to address these internal needs mounted, opportunities for moving from cold war to cooperation were eagerly seized upon by both countries. As we are finding out, this transition, like all turns to the better, is difficult and will take time. Apparently, we will have to deal with a lot of nasty little ethnic and religious wars, and sporadic acts of terrorism, before we can evolve true global cooperation, but these scattered conflicts are preferable to global nuclear war. For more on when to confront and when to cooperate, see Essay 14 and 15.

Vignette 5.2

AT SATURATION LEVELS, THE WELFARE OF THE INDIVIDUAL IS MORE IMPORTANT THAN THE QUANTITY OF REPRODUCTION.

Explanation. The importance of "lifestyle" was emphasized early in this book, especially Vignette 1.2. In the development of all natural ecosystems there comes, sooner or later, a dramatic shift from dominance by what ecologists call "R-selected" species to dominance by "K-selected" species. The "R" stands for reproduction, and the "K" represents the carrying capacity or saturation level.

We can cite goldenrods as an example of this shift in lifestyle. In an open field or meadow, a pioneer community with a lot of sunlight and unused available resources, we find species of goldenrod that allocate a large part of their energy to producing seeds, as do most plants that we call "weeds." The excess reproduction enables the pioneer, or R-selected, species, to quickly invade any "new ground" that might become available as a result of storms, plowing, or mowing. As the vegetation on the site matures and resources become all tied up in the maintenance of an increasingly crowded community, the R-selected species of goldenrod are replaced by the K types, which allocate most of their available energy to survival of the individual plant, which must now deal with very much less light and fewer minerals. Only enough seeds are produced to maintain zero growth population. As a result, flowers of the forest goldenrod are very small and inconspicuous as compared to the meadow species.

Many people are allergic to the pollen of ragweed, which is very definitely an R-selected species. What is often not recognized is that this species is not generally found in undisturbed nature but thrives best in disturbed ground, as along roadsides, recently abandoned crop fields, vacant lots, and so on. So, in part at least, ragweed is a "human-made pest."

In nature, it is very rare for any one species to be able to make the transition from R to K, that is, from the pioneer to the saturated state; usually a change or "relay" of species occurs. For example, there are no "weedy" plants (such as ragweed) in a mature forest unless an opening is created by a storm, fire, or timber cut. The challenge for humankind is to learn how to make the change in lifestyle necessary to survive in the crowded state by improving the quality of the individual and reducing the quantity of reproduction and consumption. Failing that, we will be faced with dealing with boom-and-crash cycles (as described in Chapter One) that are almost certain to occur if humankind continues to ignore limits. Then our problem will be how to recover from the overshoot without becoming violent, as happened on isolated Easter Island in the Pacific Ocean. As the once-prosperous community exceeded limits on this island, people formed warring tribes and killed off each other until there were hardly any people left. Today, all that is left is a barren grassland and the strange stone statues that stand in silent testimony to a culture that consumed the island's life-support resources and then consumed itself.

Unfortunately, we are seeing the development of this violent tribalization process in many crowded parts of the world.

CHAPTER 6

WHAT WE LEARN FROM ECOLOGY ABOUT DIVERSITY

Vignette 6.1

DON'T PUT ALL YOUR EGGS IN ONE BASKET.

Explanation. We previously commented on human proverbs, bits of wisdom coming from human experience and passed down through the generations. "Don't put all your eggs in one basket" is a familiar example. We all know that it is risky to put all your money in one stock or all your business in one product or service. Yet, we very often ignore the warning because one can often make more money specializing, at least in the short term. Over the long term, however, we usually find that our future and our fortune are safer with diversification.

And so it is in nature: diversification is evident everywhere. Over geological time, more and more species and genetic varieties have appeared, and the technology of genetic engineering has made it possible to produce more genetic varieties. Today, there are so many species on our earth that humans have not yet been able to catalogue all of them. The question is: What is the advantage of all this "biodiversity" (the term we use for this natural wealth)? And why are we concerned lately about the loss of some of this diversity due to human activities? Read on.

Vignette 6.2

REDUNDANCY IS IMPORTANT FOR RECOVERY FROM CATASTROPHIC EVENTS.

Explanation. "Variety is the spice of life," as an old saying goes, but more importantly, it enhances the long-time functioning of ecological systems by providing more than one kind of organism that is capable of carrying out a vital mission, such as photosynthesis or decomposition of dead matter. Thus, if a "keystone species" that is performing a vital function should be a victim of a disease or other catastrophe, then it is important to have one or more "backup" organisms that can do the job. This kind of functional redundancy is presumed to be a major reason why high biodiversity is considered to be important. In past ages, the earth has been subjected to comet crashes, climate changes, and other catastrophic events, but enough species in vital niches survived and enough new species have evolved to keep the whole biosphere system functioning.

We need to be especially concerned about the current lack of diversity in the varieties of food plants that are grown commercially. For example, only a handful of highly bred varieties of corn are planted worldwide. Recently there was a scare in the Midwest corn belt when a new fungus appeared that threatened to wipe out the crop. A frantic search began to find resistant varieties. As it turns out, traditional agriculture as still practiced in many less developed countries constitutes good "gene banks" because numerous varieties of food crops are cultivated. In Central America, for example, more that 20 kinds of potatoes are grown. Survival of high-yield but low-diversity industrialized agriculture may well depend on preserving lower-yield but diversified cropping techniques as a hedge against the disaster of wide-scale crop failure caused by a new pest, development of a pesticide-resistant old pest, or by climate change that adversely affects the highly bred variety. So we should not think of the age-old traditional agriculture as primitive or old-fashioned, but rather consider preserving it as part of global variety that may be important in our future.

For more on the importance of diversity at all levels of organization (genetic to biosphere), see Essays 16 and 17.

CHAPTER 7

HUMAN ECOLOGY: WHAT WE DON'T LEARN FROM NATURE

Vignette 7.1 ─────────────────────────────

COUNT ON SCIENTISTS TO RECOGNIZE PROBLEMS BUT NOT TO SOLVE THEM.

Explanation. In Vignette 1.2, we commented on the growing gap between the rich and the poor in human society. Equally distressing is the growing communication gap between science and society. The way it is supposed to work is that scientists research problems and suggest solutions that hopefully enable the public and its representatives (legislators, etc.) to better deal with societal problems. As science has become more complex and more reductionist, that is, narrowly focused, communication between the scientist and the nonscientist decision-maker has not only become increasingly difficult but also more contentious. Which brings to my mind another vignette that goes something like this: Scientists admired but not trusted by society! (See Cartoon 7.1.)

Nowhere is the communication gap more evident than in our universities. At the University of Georgia, for example, the campus is situated on two hills. On one hill, the north campus, the teaching and research facilities of the decision-making disciplines and the humanities are housed. These include economics, business, law, political science, journalism, sociology, literature, history, and so on. On the other hill, the south campus, are the basic and applied sciences, including mathematics, physics, chemistry, the biological sciences, the earth sciences, statistics, ecology, agriculture, forestry, and so on. Departmental boundaries are so tight and the requirements for majors are so rigid that it is almost impossible for a student to work out a double major between science and society as, for example, economics/ecology or earth science/political science. College administrators are well aware of this situation and are seeking means to bridge the gap, as it were, by promoting more interdisciplinary seminars, lectures, and degrees. In general, small colleges

Cartoon 7.1

do a better job of science/society education than do the large universities, because they are focused more on general education than on specialization. The overall result of excessive segregation of disciplines, not only in colleges but in the political realm as well, is a **mismatch between traditional disciplines and real-world problems**, which is to say that just about all important real-world problems by their very nature are interdisciplinary in that they require the expertise and attention of more than one traditional discipline. The anti-science attitudes we see in legislatures and the reduction in public support for basic research are in no small measure due to the difficulty specialists have in communicating to laymen their goals, which often seem far removed from the concerns of citizens. Because ecology is an integrative rather than a reductionist discipline, it can play a major role in bridging the science/society gap (which, of course, is one of my motivations for writing this book!). See also Essay 1.

Vignette 7.2

MONEY IS A VERY INCOMPLETE MEASURE OF WEALTH.

Explanation. Money is one of humankind's more important inventions and is now the basis for decision-making at most levels of society. The downside to our preoccupation with monetary values is that our money system does not take into account all the real costs and values of living. Money is not directly involved with natural wealth, which is not only the ultimate basis for human material wealth but also provides the life-supporting goods and services such as air and water recycling and purification, soil enrichment, atmospheric balances, and so on. Nor can money adequately valuate the aesthetic enjoyment of natural beauty, the arts, literature, and so on. As shown in Figure 7.1, the gross domestic product (GDP), the standard measure of economic progress, does not include social and ecological costs. In recent years an increasing GDP has not increased the "happiness index"; only 30 percent of people in the United States say they are very happy with the way things are. See also the last cartoon in the preface.

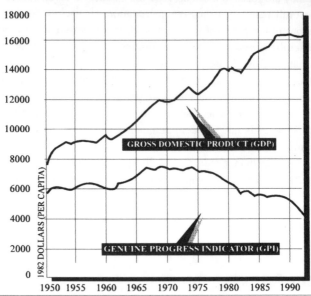

Whereas Gross Domestic Product does not account for the social and ecological costs of economic activity, the Genuine Progress Indicator does. When these costs are taken into consideration, a very different picture of the health of the economy emerges.

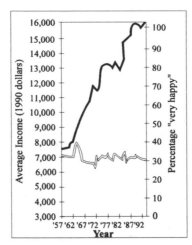

Figure 7.1 Upper: GDP is not a good indicator of economic progress because it includes economic activity associated with automobile accidents and other negative things. Lower: Feeling good. U.S. per-capita income goes up up up from 1957 to 1993. The "very happy" portion of the populace hovers around 30%.

Current market economics deal largely with human-made goods and services and very little with nature's goods and services. The so-called free market pricing system is excellent for production and distribution of human-made wealth but fails completely in sustaining the generally unpriced environmental services that are considered economic externalities. In the past, these "externalities" (air, water, and cost of waste treatment, for example) have been of little economic concern because the environment seemed large enough to absorb these costs. As human populations and their demands have skyrocketed, this assumption is no longer valid. At the present time, we rely on taxpayers to pay for these externalities (waste treatment, for example) and employ political and legal regulations to protect our life-support environmental resources. Increasingly, such regulatory efforts are too little or too late and quite rightly viewed as an unnecessary burden on the individual taxpayer. We have to find a better way to merge economics and ecology when it comes to cost accounting. For more on the shortcomings of current pricing systems, see Essay 18.

It is encouraging that both economists and ecologists are actively seeking ways to incorporate non-market values into the economic system, so that we all (businesses and the public) share in the costs and benefits. During the past decade, there has been an outpouring of books, journals, and articles, numerous conferences, and the formation of new professional societies dealing with the ecology–economics interface. Since it always takes energy to make money, one suggestion is to use energy as a currency instead of money, which could put the work of nature on a equal footing with the work of humans. One such currency championed by H.T. Odum is **eMergy** (contraction of "embodied energy"), which is the sum of all of the energy used to produce any product or service, thereby giving equal value to the goods and services of humans and nature.

Other suggestions for bridging the market–non-market gap include: (1) internalizing environmental costs so that consumers rather than taxpayers pay for waste treatment and waste reduction, and (2) charging polluters for the damage they cause.

Somehow, future capitalism must be based on the integration of nature's capital with human capital since each is ultimately dependent on the other. If governments would use tax breaks and subsidies to encourage transitions instead of always trying to maintain the status quo, I believe it will not be too difficult to develop an extended **dual capitalism** in which business, industries, and government itself will give equal attention to market forces and to internalizing environmental costs into the price of the product or service. For example, if an industry was encouraged to design its production so as to reduce its use of scarce resources (such as water or minerals) and pay for its own toxic waste treatment, free-market competition would quickly favor the industry that produces the product most efficiently with little or no toxic waste ending up in the general environment. Then the public taxpayer will not be called upon to pay for treatment that is increasingly a large part of private profits. I firmly believe that promoting incentives can greatly reduce the amount of regulation necessary to sustain environmental quality and resources. As shown in the stick-and-carrot scenario in Cartoon 7.2, a gentle use of the stick (i.e., regulation) will get the attention of industry and remind them that there is work to be done followed by liberal use of the carrot (i.e., incentives), which will encourage the development of new waste-reduction technologies. Such a strategy would involve taking a positive rather than a negative approach. For more on the ecology–economics interface, see Essays 18 and 19.

Vignette 7.3 —————————————————————————

ALL TECHNOLOGY HAS MIXED BENEFITS (another good news/bad news story).

Explanation. Just about every technological advance that is intended to improve our well-being and prosperity has its dark side. Paul Grey, an engineer by trade, former president of MIT (Massachusetts Institute of Technology), and author of a book on technology published by the U.S. National Academy Press, states that "A paradox of our time is the mixed blessings of almost every technological development." The current vignette is thus recognized as true by practical engineers as well as by theoretical ecologists!

The existence of negative side-effects does not mean that we should not continue to search for and adopt new technology; rather, we need to recognize and mitigate potential dark sides before they cancel out benefits. Two examples will serve to illustrate.

Cartoon 7.2

Modern agricultural technology involving large machines, high-yield genetic crop varieties, and heavy use of chemicals (fertilizers and pesticides) and water has increased the yield of major crops such as wheat, corn, and rice by several fold. That is the bright side. However, there are two very serious dark sides to this "green revolution": (1) water pollution by chemical and soil runoff from croplands has also increased several fold, and (2) throughout the world the small farmer who used to make a living on small acreage is being displaced by the wealthy large-scale farmer, forcing millions of people off the land and, in too many cases, into the poverty of the city slums. We might even say that industrialized agriculture has contributed to the widening gap between the rich and the poor that we discussed earlier.

Another example of mixed benefits are the large coal-burning power plants that provide relatively cheap electricity to millions, but the emissions from their tall smokestacks create "acid rain" that is devastating numerous lakes and reducing crop yields, not to mention possible negative effects on human health.

In summary, the point to be made here is that, as we seek new technologies, we must be aware that they will have dark sides that must not only be anticipated but also dealt with. Often what is needed is a **counter-technology** that will at least ameliorate the detrimental effects. In agriculture, what is known as "conservation tillage" or "residue management" is such a counter-technology that is being widely adopted. These practices not only improve the soil but also reduce the amount of chemicals needed, greatly reduce chemical runoff, and can eliminate soil erosion entirely. In the case of power plants, "clean coal" technology, in which the coal is gasified or liquefied before burning, is a proven way to eliminate acid emissions. In this technology, the acid-forming substances such as sulfur are removed from the coal before burning.

The delay in implementing counter-technologies is often due to the fact that there are **transition costs** involved in making the change. Governments should assist farmers, utilities, and industries in acquiring new equipment or whatever is needed to make environmentally favorable transitions by providing tax relief, grants, or other incentives. Political leaders who are elected for very short terms cannot be expected to face up to short-term costs that have long-term benefits unless we, the public, tell them that we want changes that will improve health and the quality of life even if there is a temporary economic cost. In the case of our agriculture and power plant examples, these short-term transition costs will be a small price to pay for sustaining the quality of air and water over the long term.

Vignette 7.4

THE TRAGEDY OF THE QUICK FIX, OR THE TYRANNY OF SMALL DECISIONS.

Explanation. Parallel with the "technological paradox" is the "quick-fix" (see Cartoon 7.3), that is, our tendency to deal with situations on a piecemeal, "one problem/one solution basis." Take, for example, a situation in which citizens are complaining about smoke from a nearby factory. A quick-fix solution very widely adopted in the past is to build the smokestack taller so that the smoke no longer descends on nearby houses. As far as the local people are concerned, the problem is solved. But, as soon discovered, the longer the stack emissions remain in the air, the more likely the sulfur, nitrogen, and other components will react with water vapor to produce deadly acids. So, in this case, the quick fix converts a local problem into national and global acid rain problems! As noted in the previous section, the only real solution is to clean up and reduce emissions at their sources, not just divert them to somewhere else.

Another related example is the 1990 Clean Air Act, which required coal-burning plants to reduce their acid-forming emissions by a third. Rather than invest in technology that would reduce emissions at their source, many plants simply switched from high-sulfur eastern coal to low-sulfur western coal. As a result, many poor miners in West Virginia were put out of work, thereby putting the burden of improved air quality on the poorest of people! The tragedy of quick-fix problem-solving is that it lulls people into thinking that the problem has been solved once and for all, and thus delays the search for more lasting solutions that require consideration of secondary and/or indirect effects. Again, we are arguing that the time has come for large "holistic" rather than small "piecemeal" approaches to human predicaments. For other examples of the tyranny of small decisions, see Essay 20.

Cartoon 7.3   Cosmetic quick fix.

Vignette 7.5

IT OFTEN PAYS TO DESIGN WITH NATURAL FORCES AND PULSES, RATHER THAN AGAINST THEM.

Explanation. Because humans have had to overcome many natural obstacles, we tend to develop a "mind-set" that we must always control nature no matter what the cost. So we build dikes, dams, and seawalls to try to control floods and storms that are a normal part of the environment. Too often such devices control small floods and storms, but when big ones come along the damage to human structures that are supposed to be protected is very much greater than would have occurred in the absence of the barriers.

A good example is building seawalls to control beach erosion. Such barriers may provide temporary storm protection for houses built too near the shore, but the beach itself is often lost because the energy of the storm tides, instead of being dissipated by flowing freely over the dunes, is reflected back on the beach, eventually washing it out (see Cartoon 7.4). Then millions have to be spent on pumping sand back onto the beach (renourishment, as it a called). A better idea would be to build houses up on tall poles so that storm tides could pass under with less damage to house and beach.

Vignette 7.6

THE THIRD ETHIC: IT SHOULD NOT ONLY BE ILLEGAL BUT ALSO UNETHICAL TO ABUSE MOTHER NATURE.

Explanation. Maintaining and improving environmental quality requires an ethical underpinning. Not only must it be against the law to abuse our life-support system, it must be understood as unethical as well. One of most widely read and cited essays on the subject of **environmental ethics** is Aldo Leopold's essay, "The Land Ethic," first published in 1933 and included in his best-selling book, *A Sand County Almanac* (1949). Leopold spent his early years as a forester in a part of the West that could only be reached by horseback and where the howl of the wolf was frequently heard. Later he pioneered the field of game management and became a professor. He did much of his writing in a cabin on an old worn-out farm in Wisconsin that he and his family restored to a place of natural beauty, a place that is now a shrine for conservationists.

Cartoon 7.4

Leopold opens "The Land Ethic" by describing how Odysseus, hero of *The Odyssey*, on returning from the Trojan War, hanged a dozen slave girls whom he suspected of misbehavior during his absence: "The hanging involved no question of propriety. The girls were property. The disposal of property was then, as now, a matter of expediency, not of right or wrong." The concepts of right and wrong were not lacking in ancient Greece, but they did not extend to slaves. In today's world, human rights, women's rights, and even animal rights receive increasing ethical as well legal attention. What about the environment? Leopold suggested that an extension of ethics over time is like the following sequence. First, there is a development of religion as a human-to-human ethic. Then comes democracy as a human-to-society ethic. Finally, there is the yet to be developed ethical relationship between humans and their life-supporting environment. In Leopold's words, "the land-relation is still strictly economic, entailing privileges but not obligations."

Another way to show the importance of developing the third ethic, as it were, is to project two divergent futures or "scenarios" for humankind, something like as what follows:

1. If ethics and law are restricted to survival of the human individual, as is mostly the case now, then we will have uncontrolled fertility and exploding human populations that, combined with neglected and degraded life-support ecosystems, will result, at best, in a miserable future for humankind.

2. If, instead, ethics and law are extended to the survival of species (ours and others that we depend on) and on maintaining the quality of the environment, human population overshoot growth will be brought under control, which, combined with healthy ecosystems, will result in a quality future for humankind.

Vignette 7.7

DOMINION AND STEWARDSHIP: A TIME SEQUENCE, NOT A CONTROVERSY.

Explanation. There is an interesting parallel to the youth–maturity theme, as discussed in Vignette 4.2, in the scriptures. Early on in the Bible we are told to be fruitful, multiply, and take dominion over the earth (Genesis 1:28). One interpretation of this message is that it was meant to apply to all living things (all of God's creatures, as it were), not just humans. Elsewhere in the scriptures (Luke 12:48; I Corinthians; Psalms 37:27–28, among others), we are admonished to become "stewards" (a word that comes from a root meaning "keeper of the house") and to take care of the earth. A reasoned interpretation is that these messages are not contradictory, nor a matter of right and wrong, but constitute a sequence in time, which is to say that there are a time and place for each.

In the early stages of civilization, taking dominion over the environment and exploiting resources (clearing the land for crops, mining the earth for materials and energy, etc.) as well as high birth rates were necessary for human survival. As our society becomes crowded, ever more resource-demanding, and technologically complex, there is less need for large families and child labor, and, more important, various limitations are reached that force us to think and act as stewards. It is now very much in our interest as individuals to begin to do so. For more on these ideas, see Essay 25.

CHAPTER 8

BOTTOM LINES

I have tried to show with "sound bite" vignettes, cartoons, and charts how ecological thinking and human common sense can help us understand and deal not only with environmental problems but with other human predicaments as well. In fact, the environment is no longer an issue separate from such human concerns as crime, welfare, health care, food, or pollution but is an important part of all of our problems and aspirations. Perhaps for the first time in human history we are faced with the fact that too much of a good thing (material wealth, consumption, automobiles, technologies, even people) is becoming as great a threat to our future as too little. Bigger may no longer be better, and quality is certainly better than quantity over the long run, so we have to modify or extend our economic system to include all of the values that make for quality of life. To do this requires some major changes in the way we think, educate, and do business.

We can list at least five major changes that must be made. We have already covered four of these in previous chapters, namely: (1) replacing boom-and-bust growth with sigmoid growth (quality over quantity), (2) incorporating non-market (life-support) goods and services into the economic system (dual capitalism), (3) getting really serious about land-use planning, and (4) more cooperation and less confrontation. A fifth need is to greatly reduce or eliminate toxic wastes, as noted in Vignette 7.4.

As shown in Cartoon 8.1, it is getting hard to find a place to dispose of wastes that is not in someone's backyard. On a recent visit with a middle school class, I projected this cartoon on the screen and asked the class if there was a message here. Several hands went up with answers to the effect that there is no longer any place to dump the stuff. Then I asked, "What are we going to have to do?" There was a longer pause, and a student in the back of the room suggested, "We will just have to quit producing the stuff." So, there you have it. The wisdom of a child: SOURCE REDUCTION AS THE ULTIMATE SOLUTION TO POLLUTION. In Vignette 7.2 we noted that wasteless industries might become a reality with carrot-and-stick cooperation between industry and government.

"DON'T YOU KNOW? WE JUST DRIVE AROUND. THIS IS A MOBILE TOXIC WASTE DUMP."

Cartoon 8.1

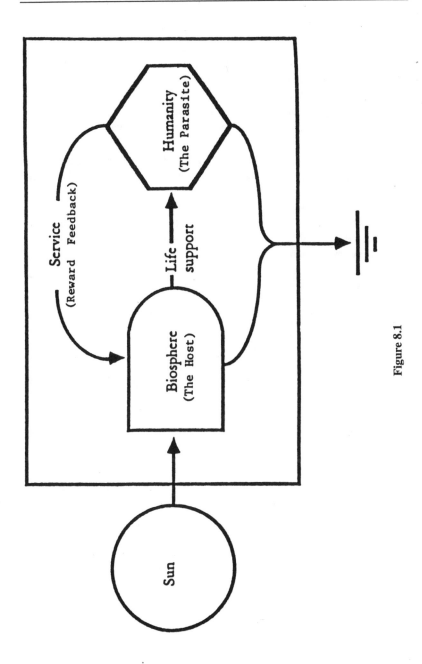

Figure 8.1

Figure 8.1 sums it all up. We must begin to devote more of our human wealth, energy, and engineering skills to servicing and repairing our "big house," the biosphere, which provides not only a place to live and enjoy but also all of our life-support needs. In a very real sense, humanity is a parasite on the biosphere, since as a non-green consumer organism we depend among other things on green plants for food and oxygen and on microorganisms to recycle nutrients. As we learn from studying parasite–host relations in nature, a prudent parasite that has only one host does not kill off that host, since that will result in its extinction. The prudent parasite moderates its demands on the host, and in many cases actually does things to help its host prosper (the "reward feedback" concept mentioned in Chapter Four). In ecological language this is **co-evolution for co-existence**. We cannot for much longer put off repairing and servicing our overcrowded and deteriorating house and putting some reasonable limits on its future use and occupancy. To do all this will require more environmental engineering technology. Most of all, it demands more cooperation and less confrontation, both within the human population and between the human population and nature.

ESSAYS AND COMMENTARIES
FOR FURTHER READING

Harmony Between Man and Nature: An Ecological View

Eugene P. Odum

O ur generation enjoys the fruits of technological achievements that have increased energy and resource availability far beyond subsistence. But at the same time we are confronted with exploding populations, dwindling natural landscapes, and mounting pollution, which threaten to cancel many of the advantages of this energy-rich society. The realization that man is an integral part of his surroundings has directed our attention, more and more, to understanding and protecting our endangered environment and to reordering technology to achieve new goals. Thus, interest in ecology and application of ecological principles to human affairs has become a new intellectual frontier. Environmental awareness, worldwide in scope, reflects that, for the first time in its history, humanity faces ultimate rather than merely local limits to human needs and aspirations.

In revolutionary times such as these, it is to be expected that many long-accepted concepts and attitudes relating to man and his environment should be brought into question. Intense analyses and debate on issues are to be encouraged, as is the establishment of communication between vested interests within our society. This is a necessary step in making decisions for the good of the whole. Controversy is equally important at the professional

Reprinted from: Beyond growth: essays on alternative futures. *Yale Univ. School of Forestry Bull.* **88**:43–55, 1975.

level because such specialists, in emphasizing precision of communication *among* themselves, pay the cost of lessened communication with other professional groups, or the public at large. Each profession has its own "articles of faith." These need to be reexamined periodically to make certain that they are consistent with new knowledge and changed conditions. In this discussion, I shall make a special effort to *broaden* the concepts in the hope that all can understand them. I am optimistic that the basic intelligence of man will surface in time to implement the orderly reforms that must be made to meet environmental contingencies, provided controversies do not become refractory and there are not too many false starts and backlashes that generate dangerous oscillations and uncertainties.

One long-accepted concept, now being challenged, is the belief that economic and population growth are necessary to raise the standard of living and cure societal malfunctions associated with pollution, hunger, crime, and unemployment. Writing in a current symposium volume entitled *Energy, Economic Growth, and the Environment* (Schurr, 1972), economists Walter Heller and Kenneth Boulding present contrasting viewpoints. Heller says, "One rightly views growth as a necessary condition for social advance, for improving the quality of the *total* environment." Boulding says, "One might even have an optimistic image of the present period of human expansion as a kind of adolescence of the human race in which man has to devote a large portion of his energy to sheer physical growth. Hence we could regard the stationary state as a kind of maturity in which physical growth is no longer necessary and in which, therefore, human energies can be devoted to qualitative growth — knowledge, spirit and love."

As an ecologist with a holistic view of man and nature, I would agree that Boulding's youth → adolescence → maturity sequence is a useful, but, by itself, an inadequate analogy for comparing development at different levels of organization. Almost always what is true at one level explains only part of what is true at other levels — this being the well-known principle of "integrative levels" (see Fiebleman, 1954). To understand a process, such as development, at a particular level we must consider the process *at that level*. Although there are parallels between the individual and society in regard to growth and maturity, there are additional "scale factors" to be considered when we move from individual levels to the level of society, and especially when we consider man and his environment as an ecosystem.

My principal theme is as follows: Development at the ecosystem level differs from development at the individual level in that *aging and death do not inevitably follow achievement of maturity*, as is the case in the individual organism. I sincerely believe the reason many people become highly emotional on the subject of "Limits of Growth" is that they fear that if the physical growth of society stops it will then deteriorate and die. Taking cues from what we can observe about the development of large systems, I believe we can show that this is by no means the case. In fact, the truth may be just the opposite. If the physical growth of society does not level off at an optimum size in terms of the resources and life-support system on which it depends, the *continued improvement of the quality of human life will be more and more threatened by cancerous growth* — uncoordinated and uncontrolled parasitic growth that becomes lethal when it can no longer be supported by the system of which it is a dependent part.

In the ecological context, the important questions do not relate to limits of growth per se, and certainly they should not focus on making a choice between the extremes of "zero growth" and "laissez faire (i.e., uncontrolled) growth." Rather, after sheer physical growth in size becomes undesirable or impossible, the focus should be on extension in the growth of the quality of human existence. It is the "law" of an ecological system that individuals or components within the system die and are replaced by perhaps better individuals. This makes it possible for the whole system to improve rather than to age and die. Likewise, if society is not to age and die, economic and social institutions must also evolve (i.e., be replaced by more adapted ones) as the stages in development change. To blindly adhere to social and economic institutions based on the concept that "bigger is always better," just because this is true during the early stages of development, would be to invite the death of the system, because it prevents necessary movement to the next stage. To rephrase this viewpoint, we could say that the problem society faces is to make an orderly transition from a youthful stage, in which man's relationship to the environment must of necessity be exploitative and parasitic, to a more mature stage of harmony between man and nature. We can find interesting models of this transition in natural systems, where we observe that parasitism tends to evolve into mutualism when components live together over long periods of time in a stable system. Any parasite that does not make this transition risks extinction by destroying its host (for additional discussion of such models see Odum, 1971, pp. 220–233).

In our society, with its specialists and special interests, the word "growth" means different things to different people. Many businessmen, economists, and planners are preoccupied with growth in size, while the ecologist recognizes many other kinds of growth that do not involve a net increase in system size or in the number of components. For example, in a forest developing on an abandoned field, through time there will be growth in efficiency of use of energy and resources and in recycling of materials; there will be growth in

1. Increases in efficiency in the use of energy.

2. Increases in efficiency in the use of resources.

3. Increases in turnover rate of components (repair before deterioration occurs).

4. Increase in rate and efficiency of recycling of materials.

5. Increase in proportion of energy devoted to maintenance.

6. Increase in diversity of components.

7. Increase in stratification.

8. Increase in quality of components (continual improvement of existing superior components rather than proliferation of more inferior ones).

It is interesting to note that, according to one theory, aging in the individual comes when cycling cells (components) become non-cycling (see Gelfant and Smith, 1972, for a review of this theory). Whether this is a useful analogy for the ecosystem yet remains to be determined.

Table 1.1   Types of growth that do not require a net increase in size but that can function to prevent aging, not only in the mature (steady-state) natural ecosystem, but also in a mature society

diversity of organisms, in stratification, and in the size and quality of individuals. All of these kinds of growth, which do not require increase in size, function to prevent "aging" of the ecosystem. There is no reason why "substitutes" for growth in size cannot generate employment, GNP, and higher standards of living in the equilibrium society just as effectively (or maybe even more effectively) as does growth in size in the pioneer society. Refining, rebuilding, recycling, and servicing already developed systems to make them more efficient can certainly furnish plenty of opportunity. In Table 1.1 are listed eight types of growth that not only appear to be important in the natural mature ecosystem, but that could theoretically also play a role in preventing aging in the total man–environment system. Poor people and culturally disadvantaged people need not fear that phasing into a mature society will cut off their opportunities for individual development. Economists and social scientists often discuss some possibilities along these lines, and I shall draw a few parallels myself at the end of this presentation.

While most thoughtful people will probably agree with the overall logic of my argument, there is still a widespread belief that we have not yet reached any real limits for physical growth, or that technology can somehow extend the limits indefinitely. Since energy is the common denominator for man and nature, let us analyze the role of energy in development of society and try to determine if: (a) the time has come for man to plan for a transition from quantitative to qualitative growth, or (b) we can continue to enjoy youthful growth for some time to come and let some future generation worry about the mature state. There are five basic principles for such an analysis. These are:

1. In a technologically advanced industrial society, energy itself is not likely to be limiting, but the consequences of converting energy from one form to another is limiting. When we run out of fossil fuel there is plenty of potential energy in the atom, but tapping such new sources will be more difficult. The technology will be more sophisticated and its maintenance and repair more difficult and costly. In a recent case in the New York area, seven months were required to repair a malfunction in the heat-generating component of an atomic power plant; only a couple of days would have been required to repair a similar breakdown in a coal-fired plant. Accordingly, it is a challenge to keep the unit cost of a kilowatt or BTU of energy from rising. In fact, unit costs are rising and will rise despite our best efforts. The relatively cheap power rates of the past decade, responsible for so much recent rapid industrial development, resulted both from the relative ease of converting fossil fuels to useful work, and the fact that we have postponed paying some of the costs, especially environmental costs. As the cost of energy conversion increases and the space for environmental waste treatment decreases, the *total cost* (not just dollars and cents cost) of energy will be driven higher, which means that its conversion from one form to another must be much more prudently managed than is now the case. At the present time, as shown in Table 1.2, only about half of all the energy used in the United States is converted to useful work; the other half is waste (i.e., pollution). Automobiles and electricity are very convenient and important in generating economic wealth, but as presently designed both are particularly inefficient in energy conversion (only about 30%) and generate mountains of waste that strain our life-support systems. Increasing efficiency, of course, is one way to provide more energy to people who do not now have enough without increasing total consumption. As already mentioned, this is one of the growth substitutes utilized by mature ecosystems.

2. There is no technological way to bypass the second law of thermodynamics, which states that as energy is used a proportion is converted or

Type	Gross consumption	Consumed as		Useful work accomplished			Totals	
		Electricity	Fuel	Household & commercial	Transportation	Industrial	Useful work	Waste
Nuclear	0.05							
Hydropower	0.68							
Gas	6.13	4.28 +	12.0	3.18 +	1.03 +	4.06	8.27	8.01
Petroleum	6.02	(26%)	(74%)	(38%)	(13%)	(49%)	(51%)	(49%)
Coal	3.40							
Total	16.28	16.28		8.27			16.28	

Table 1.2   U.S. consumption, use, and waste of energy — 1970

dispersed in an unusable (by man) state; energy cannot be recycled as can materials. One can ameliorate thermodynamic disorder created by energy conversion, but one cannot avoid the basic cost of dealing with it. Concentrating and intensifying the use of energy within small areas (cities, heavy industries, etc.) creates especially difficult thermodynamic disorder problems. It is evident that disorder problems (various forms of pollution) have already resulted in diminishing economic returns and in leveling growth in parts of the United States. Even our efforts to increase food production in fuel-subsidized agriculture (where large amounts of fossil fuel are used to raise agricultural output) are beset by increasing thermodynamic disorder that will eventually increase the cost per unit of food produced. To double plant food production, a tenfold increase is required of fertilizers, pesticides, and horsepower — all of which pollute (see Odum, 1971, p. 412, for documentation of this relationship). Modern methods of converting grain into meat bring additional losses and costs, as evidenced by water pollution caused by animal feedlots. We have scarcely begun to think about paying for the treatment of agricultural waste. In contrast to most countries, agriculture in the United States now consumes more fuel energy calories than it produces in food calories.

3. Exponential growth cannot continue for very long. We have two options: (1) let positive feedback run its course until overshoot occurs (boom and bust), or (2) install some negative feedback to control growth, to prevent overdevelopment, and to buy necessary time for solving problems. Lost in the controversy between the extremes of "zero-growth" and "laissez-faire growth" is the simple, common-sense truth that a controlled, moderate rate of growth not only prolongs the time when we can enjoy the undeniable benefits of growth but also reduces the cultural and technological lags that are inherent in rapid growth.

4. As the size of a system increases, the cost of maintenance of a network of services increases as some kind of power function, at least as a square, viz:

$$C = \frac{N(N-1)}{2}, \quad \text{which approximates} \quad \frac{N^2}{2},$$

where N is the number of units in a network and C is the maintenance cost. Thus, doubling the size of a city or a power plant is likely to more than double the cost of maintenance. Since there are other "tradeoff" advantages to an increase in size (i.e., "economics of scale" factors), bigger is better up to a point. Theoretically, there is an optimum power concentration per unit area beyond which any increase in size costs more than it's worth. In other words, cost–benefit curves are really humpbacked and not the straight lines all too often projected by "superficial optimists." Figure 1.1 shows a generalized information input–output performance curve that has been shown to apply to several levels of biological organization. Presumably, some such curve applies to power-in/benefit-out relationships. One insidious feature of such a curve is that the downward trend that comes after the optimum is reached is so much more gradual than the pre-optimum upward trend. Thus, increases in output per unit input benefits are quickly recognized in the early stages of development, but declining benefits after the optimum has been passed are not so easily recognized. Unfortunately, large, power-hungry, man-made systems have an inherent tendency to overdevelop, since profits can be made by going beyond the optimum so long as payment for the increased cost of maintenance is avoided or delayed. In my opinion, locating the optimum plateau in the energy–performance curve is the greatest challenge in systems research today. Fortunately, most factions in our political system are beginning to talk about economic reforms, especially in the area of taxation, designed to counter the overdevelopment syndrome.

5. The natural environment, the life-support system for both man and his energy-consuming machines, is the major means for tertiary waste treatment. The work of nature in pollution abatement has been underevaluated, because

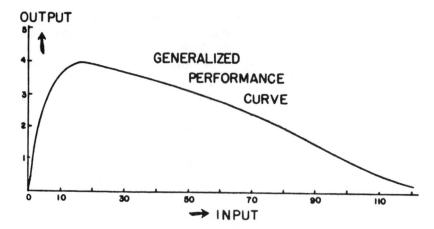

Figure 1.1 Theoretical curve based on performance data of biological systems under various rates of information input (figures are
in relative information units — bitssecond × 10^3; by "output" is
meant proportional output per unit of input)

this service has been *free*. Present cost/benefit procedures generally fail to
include this basic dependence on the natural environment until *after* the
overshoot, i.e., until the environment is badly polluted and suddenly we have
to pay for treatment that was previously free. In terms of energy flow, there
are three basic kinds of ecosystems that must function together in some kind
of balance if man is to prosper. Natural systems operate on dilute sun energy
and, therefore, are low-powered. At the other extreme, cities and industrial
complexes are subsidized by huge fuel imports (see Table 1.3) and, therefore,
operate at a power level three orders of magnitude higher than those natural
systems. However, nature is very efficient at its level of power when one
considers that natural systems are self-maintaining and (by and large) take
care of their own wastes — in contrast to cities, which, as we now manage
them, are almost completely parasitic on type 1 and 2 systems, as listed in
Table 1.4. Furthermore, nature does its work without economic cost to man,
so long as its capacity is not exceeded. Because of the low unit area capacity,
very large areas of the natural environment are required to absorb the
thermodynamic disorder produced by the high-energy systems of man. When
such environment is not available, or is so severely stressed that it is unable
to maintain itself and also do the extra work of pollution abatement, then

	Power requirements (kcal/m²/year)	Energy flow (kcal/m²/year)
1. Natural solar-powered, unsubsidized ecosystems[a] (these are the basic autotrophic, self-maintaining life-support systems	1,000 (av. 2,000)	10,000
2. Solar-powered ecosystems subsidized by free natural[b] or man-controlled, but expensive, inputs[c] of energy and materials (these are the basic food-producing systems)	10,000 (av. 20,000)	40,000
3. Fuel-powered urban-industrial systems (these are the wealth-, and pollution-, generating systems that are parasitic on ecosystem types 1 and 2	1,000,000 (av. 2,000,000)	3,000,000

[a]Oceans, natural forests, and grasslands are examples.

[b]An estuary is an example of a natural ecosystem subsidized (and its productivity thereby enhanced) by natural tidal energy.

[c]Industrial agriculture is an example of a solar-powered ecosystem heavily subsidized by fossil or other fuel at appreciable cost to man.

Table 1.3 Ecosystems classified according to source and level of energy

artificial tertiary treatment of large volumes of low level wastes (CO_2, heat, NO_2, SO_2, tritium, etc.) becomes necessary. In the long run, it will be these low-level/large-volume wastes, together with the outright poisons (lead, mercury, DDT, industrial phenols, etc.) that pose the greatest threat to the quality of human life. If all these wastes had to be treated artificially, costs would skyrocket!

Based on two different models, my brother and I (see Odum and Odum, 1972) have suggested that, in a state or region with one or more large industrial cities, at least 50% of the land and shallow waters should remain in a natural state if quality, in terms of human health, nutrition, and use options of both city and country, is to be maintained. Whatever proves to be an optimum mix of developed and undeveloped environment, one thing is certain: the former will overshoot the latter unless comprehensive regional land-use and energy management plans are adopted — soon.

When these five basic principles are considered, it is evident: (1) *that for the first time in his history man does indeed face ultimate, rather than merely local, limitations, not so much in terms of energy sources but rather in terms of impacts created by the concentration and conversion of energy*, and (2) *that environment is rapidly becoming the factor that sets the limit for man's fuel-powered developments*, and, in turn, for the number of people that can be supported at a reasonably good standard of living.

The Forrester–Meadows models (see Forester, 1971; Meadows et al., 1972) show that a disastrous overshoot *could* occur between the years 2000 and 2100 if present day economic and population growth and resource-wasting technology continues unmodified. The great value of their models is the easily understood warning of what could happen. A logical response is to examine the parameters in the model to see which ones need to be changed to prevent an overshoot. My review of the five basic principles suggests that reducing the economic pressures that push the development of fuel-powered systems beyond the capacity of the life-support systems would reduce chances of an overshoot. Likewise, a general shift from quantitative to qualitative growth would greatly change the output of the Forrester–Meadows model.

While the specific environmental models briefly referred to in this paper are not yet firm enough to provide a basis for broad-scale environmental-use planning, they do indicate that all the oceans and other large bodies of water, plus perhaps as much as 4 to 5 acres of solar-powered terrestrial natural, or semi-natural, environment *per person* must be kept in good working order as tertiary treatment systems if the quality of urban industrial areas is to be maintained. Since the density of the United States, and also the world as a whole, is now about one person for every ten acres of land area, we have time and opportunity to develop plans for maturation at an optimum rather than a maximum level of energy flow.

It is clear, at least to an ecologist, that new strategies are necessary when society moves into the mature state. I say "when" because some regions are yet undeveloped and need quantitative growth, while others have already passed the optimum. In an overall sense, reforms in two major human institutions, economics and planning, are needed *now* so that the transition to maturity can be made as needed without first suffering dangerous over-shoots.

For example, present economic and political systems tend to: (1) favor the wastemaker over the public interest, (2) stimulate an artificial demand for hard goods, and (3) encourage a wasteful technology that converts degradable natural resources into nondegradable and poisonous byproducts. It should not be too difficult to alter tax procedures, which now stimulate these responses, so as to discourage them.

In the area of planning, we must move toward regional environmental-use planning. As an indication of how this might be accomplished, I call your attention to the recently published *California Tomorrow Plan* (Heller, 1971). It is remarkable that this little book that outlines two alternate plans for the state was the work of a citizens' group rather than professional planners, although, of course, the expertise of planners was drawn upon extensively. One plan, called "California One," projects what California would be like in the year 2000 if present approaches to problem solving continue without change. The present disruptive forces would, in all probability, intensify. The "California Two" scenario has government and private enterprise join forces in a well-financed state commission to design a statewide land-use plan. The plan, amended by open discussion, would be voted into effect by the public. It is interesting to note that the projected land-use plan calls for large areas (perhaps 40% of the state) to be in "conservation" and "regional reserve" categories (i.e., preserved in natural and semi-natural states). While California Two is no utopia, it is projected that many disruptive tendencies, such as crime, pollution, environmental decay, and social disorder, would be reduced rather than intensified. Further, California Two is shown to be politically and economically feasible, requiring no change in our form of government. Perhaps even more important, coordinated regional control of environmental quality is thought to reduce, rather than increase, centralized (i.e., federal) control of the individual's options regarding such things as housing, health, education, and recreation. These matters would be more under local control with a strong regional plan in operation. This is important, because many people now reject the idea of large-scale planning in the belief that it reduces individual liberty and brings on more bureaucratic control. According to Heller's little book, this need not be the case.

Faced with a choice between California One and California Two, most people would certainly be inclined to choose the latter option. In any event, I believe we have here the first tentative step in the direction of putting into practice the theory of "ecosystem management."

Optimum Population and Environment: A Georgian Microcosm

Eugene P. Odum

The world seems to be getting smaller and more limited in its capacity to support human beings because the per-capita use of resources in developed countries and the per capita expectations in undeveloped countries keep going up. Man must now manage his own population as well as the natural resources on which he depends.

To the ecologist, this means that the population growth rate must be drastically reduced so that an equilibrium can be reached. If this is indeed the case, then the question of what constitutes an optimum population density for man becomes a key issue. An ecological approach to this problem involves considering the total demands that an individual makes on his environment, and how these demands can be met without degrading or destroying his living space, or "lebensraum."

Since the environment is both a supply depot and a house for man, the concept of the integrated system, the ecosystem, is the basis for the relevant ecology of today. In the conduct of human affairs in the past, these two functions of the environment have been considered as separate and unrelated problems. The dramatic change in people's attitudes toward their environ-

---

Reprinted from *Current History*, **58**:355–359, 1970. Also published in *The Ecologist* **1**(9):14–15, 1971.

ment and the rise of a sort of "populist" ecology in the 1970s stem from a general recognition of the fact that the quality of the lebensraum is so intimately interrelated with the rate of production and consumption that beyond a certain level an increase in the latter seems to cause a decrease in the former.

The Georgian Microcosm

In the fall of 1969, my class in advanced ecology at the University of Georgia tackled the question of the optimum population for Georgia on the assumption that this state was large enough and typical enough to be a microcosm for the nation and the world. The following question asked was. How many people can Georgia support at a reasonably high standard of living? We found that Georgia is a good microcosm for the United States because its present density and growth rate, and the distribution of its human and domestic animal populations, are close to the mean for the whole nation. Likewise, food production and land-use patterns in Georgia are average. Furthermore, since pollution, overcrowding, and loss of nonrenewable resources have not yet reached very serious proportions, the state, like most of the nation, has the opportunity to plan ahead for a new kind of progress, based on the right of the individual to have a quality environment and to share in the economic benefits of wise use and recycling of resources.

As background for the Georgia inventory, two general principles were adopted. The first: The optimum is almost always less than the maximum. In terms of human population density, the number of people in a given area that would be optimum from the standpoint of the quality of the individual's life and his environment is considerably smaller than the maximum number of people that might be supported, that is, merely fed, housed, and clothed as dehumanized robots or domestic animals. The second principle: Affluence actually reduces the number of people who can be supported by a given resource base. The per-capita consumption of resources and the production of wastes are much greater in the developed countries. If one person in the United States exerts fifty times more demand on his environment than does one Asian, then it is obvious that no environment can support as many Americans as Asians without disastrous deterioration in the quality of that environment. Shocking as it may seem, the United States is now in as much danger of overpopulation at its level of per-capita living as is India at her present standard of living. Population control must be an overriding issue in both the developed and underdeveloped worlds, but the levels that are critical, the limiting factors and the strategy of control are quite different. In underdeveloped countries the standard of living for the individual must be raised before population control can be effected; in developed nations the wastage of resources caused by overemphasis on personal consumerism will have to be reduced if population control is to do any good.

Minimum American Per-Capita Acreage Requirements

Figure 2.1 shows an estimate made by students of the minimum acreage necessary to support one person at a standard of living now enjoyed by Americans. This included a pollution-free living space, room for outdoor recreation, and adequate biological capacity to recycle air, water, and other vital resources. The per-capita area required for food was obtained by taking the diet recommended by the President's Council on Physical Fitness and determining how much crop and grazing land is required to supply the annual requirement for each item. If Americans would be satisfied with merely getting enough calories and greatly reducing their consumption of meat, as little as a third of an acre per person would be adequate. But the kind of diet Americans now enjoy — including orange juice and bacon and eggs for breakfast, and steaks for dinner, all of which require a great deal of land to produce — takes at least 1.5 acres per capita. Thus, the American demands from his agricultural environment ten times the space that is required to produce the rice diet of the Oriental, assuming equally efficient crop production in both cases. The 1-acre requirement for fibers is based on present per-capita use of paper, wood, cotton, and so forth, that equals the average annual production of 1 acre of forest and other fiber-producing land. The 2 acres for natural area use are based on the minimum space needs for watersheds, airsheds, green belt zones in urban areas, and recreation areas (public parks, golf courses, fishing and hunting areas, and so on) as estimated by recent land-use surveys. Again, we could do with less by designing more artificial waste recycling systems and doing away with outdoor recreation, but at a high cost to society as a whole.

In considering the 5-acre per-capita estimate, two points must be emphasized. If per-capita use goes up in the future, either more land is needed or greater production per acre must be forced by increased use of chemical controls. These controls, in turn, tend to pollute the total environment,

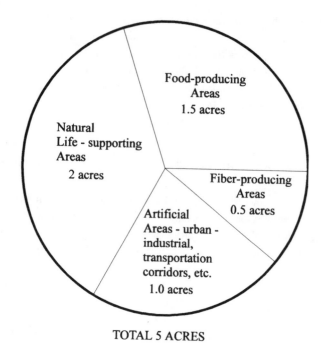

TOTAL 5 ACRES

Figure 2.1   Minimum per-capita requirements for a quality environment.

creating a cost in taxes that would reduce the individual's take-home pay. The 5-acre estimate is relevant only to an area such as Georgia that has a favorable climate with adequate rainfall and moderate temperatures. The per-capita area requirement would be much greater in regions with large areas of deserts, steep mountains, or other extreme ecosystems.

If we assume for the moment that one person in 5 acres is a reasonable per-capita density, then Georgia is rapidly approaching that level. The net growth rate is 2%, which, if continued, would mean a doubling of the population in 35 years, leaving only 4 acres per capita. Almost before we realize it, Georgia will be moving from what was considered to be a sparsely populated state to one beginning to feel the adverse effects of population pressure. As emphasized, this pressure is due not so much to the number of people, but to the great increase in the per-capita demands on space and resources. It comes as a shock to everyone that Georgia and the nation could be badly overpopulated by the year 2000.

Natural Regulators

Animal populations in nature normally regulate their density well below the
limit that would be imposed by the food supply. In this event, the quality of
both the individual and the environment is insured, since the individual is
likely neither to run out of food, or other resources, nor to overgraze or
otherwise permanently damage his habitat in his efforts to obtain the neces-
sities of life. In some populations, death, brought about by predators, disease,
or parasites, is the regulator; in other populations, birth control is the mechanism. In some of the best-regulated species of the most highly evolved
animals, namely the birds and the mammals, the essential control is behavior
that restricts the use of space.

This sort of "territorial control" would seem to be relevant to the human
population problem. Best of all, planned and controlled land use mutually
agreed upon through the democratic process can be accomplished at the local
and state levels right now, while we continue discussions about birth control
and abortion in an effort to reach some kind of national and international
consensus that can make these approaches effective nationwide and worldwide. Consequently, it certainly will be worthwhile to consider what we
might accomplish along the lines of territorial control through land-use
planning.

Land-Use Planning

In actual fact, Georgia is extremely vulnerable to overpopulation for two
reasons: (1) the immigration rate is high and can be expected to increase as
people flee from the crowded, polluted, and deteriorated part of our country;
and (2) land is open to immediate exploitation on a huge scale because there
are so few protective laws and so little land in public ownership. Many of
these factors apply to other areas of the nation. Even if the birth rate drops in
Georgia and other less crowded states, population growth rate would remain
high because of immigration that will come as people discover the relatively
cheap and quickly available open spaces. As already indicated, a growth rate
of 2% per year means that Georgians would be down to one man in 4 acres
within 35 years.

A land speculation spiral that is economically ruinous to all but a few
speculators could well result unless plans are made now, and control legislation is enacted. Georgia has a lot of open land now, but very little has been
set aside to remain so. Only about 7% of Georgia is reserved in national,

state, or city parks, refuges, greenbelts, or other protected land categories; even the best farmland is vulnerable to real estate exploitation.

What can citizens in Georgia, or in any other state where such problems exist, do? First, they can instigate and support drives, both at the local and state levels, to get more land into public ownership (parks, state and national forests, greenbelts) and can work to have an "open space" bill passed that will enable private owners to establish scenic easements and other restrictions on the use of land that is valuable in its natural state. Second, they can work towards the establishment of metro commissions and statewide environmental commissions with strong zoning powers. For example, the passage by the Georgia legislature of a marshlands protection bill early in 1970 was a step in this direction. Almost half a million acres were put into a protective category with an agency empowered to ensure the best and highest use of a natural resource that otherwise is very vulnerable to destructive types of exploitation.

If about one-third of the area of Georgia were in a protected category, then it would be well protected against overpopulation, and would have a big buffer that would make the technical problems of pollution control much easier. It is important to note that Western states are fortunate in that 40 to 50% of their land is already in public ownership. The battle there will be to mobilize public opinion to prevent overdevelopment and degradation of these lands.

The third function that citizens can perform is to be more selective about the type and location of new industry they can allow in their states. Citizens will be doing industry and society a favor by establishing tough pollution standards, and requiring advanced waste treatment. It is much cheaper to engineer and internalize the costs of complete waste treatment and water-and-air recycling at the beginning than to take action later and also pay for repairing a damaged environment. There is no longer a need or excuse for "dirty" industries that pollute and pay low wages. Any state can now attract industries that have the resources to pay good wages and the public conscience to do what is necessary in waste management.

In summary, our microcosm study makes a case for basing the optimum population on total space requirement and not on food as such. The world can feed more "warm bodies" than it can support high-quality human beings.

The World's Most Polymorphic Species: Carrying Capacity Transgressed Two Ways

William R. Catton Jr.

*B*iology is, as Hardin (1986) has reminded us, rich with insights that indicate a need for a "massive restructuring of popular opinions." In particular, the supposition that earth is a cornucopia for mankind needs serious modification. Unfortunately, appropriate opinion restructuring is impeded by an inherent antagonism: although ecologists recognize there are limits to ecosystem sustainability, politicians are professionally compelled to remain deaf to suggestions that growth of human activities and elevation of consumption cannot be perpetual. The ecologists' time horizons are based on evolution or succession; politicians' horizons are seldom more than two or four years away, because they get reelected by encouraging electorates to expect them (at least in election years) to promote economic growth.

This article will offer suggestions for getting some fundamental ecological insights onto the public policy agenda. Specifically, I will try to go a step beyond Hardin and make the case for remarriage of sociology and biology as a means to this end. Although knowledge of both the ecological and

Reprinted with permission from *Bioscience* **37**:413–419, © 1987 American Institute of Biological Sciences.

sociological nature of the human species is politically necessary to forestall disaster, few national leaders yet recognize it. Carrying capacity needs to be understood as the maximum load an environment can permanently support (i.e., without reduction of its ability to support future generations), with load referring not just to the *number* of users of an environment but to the total demands they make upon it. For human societies, as for populations of other species, the relation of load to carrying capacity is crucial in shaping our future. Public comprehension of the concepts of carrying capacity and load is both vague and inadequate, and the need to correct these deficiencies is urgent.

Human Ecology

For two reasons, *Homo sapiens* is a species especially likely to transgress an environment's sustainable carrying capacity. First, humans have an unusually long period of maturation. Therefore, sociologists commonly view learned culture (rather than biological instincts) as the explanatory mainspring in accounting for human behavioral patterns. We must now also see that the modesty of an infant's demands upon an ecosystem obscures the immensity of the load each adult may later impose. Second, our cultural nature enables our wants vastly to exceed mere physiological appetites.

Ecology has long been described as the study of interrelationships among organisms and their environment. My task is to show what special turns in such study are required by the special nature of human organisms. When sociology shuns such biological concepts as carrying capacity (or distorts their meaning in embracing them), it ignores important determinants of human experience. But unless ecologists take the facts of human culture appropriately into account, they, just as truly, are being unrealistic. Various sociologists who have styled themselves human ecologists have purported to resolve the differences. Let us examine some of their efforts.

Aware of the central importance of the ecosystem concept, Duncan (1959, 1961) sought to adapt that concept for use in human ecology, taking into account two fundamental ways in which humans differ from other organisms in their ways of environmental interaction. Humans develop technology, and humans organize into groups more elaborately and more variably than nonhuman populations do. So to Duncan the human ecological version of the ecosystem concept seemed to consist of what he called an ecological complex comprising four classes of interdependent variables: population, organization, environment, and technology (POET).

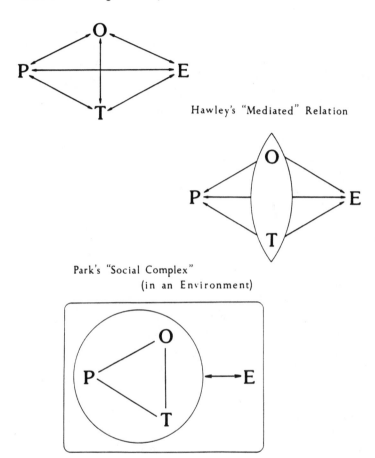

Duncan's "Ecological Complex"

Hawley's "Mediated" Relation

Park's "Social Complex"
(in an Environment)

Figure 3.1 Three views of human ecology involving population (P), organization (O), environment (E), and technology (T).

Hawley (1973), reacting somewhat sternly to a paper by Odum (1969), also insisted on the special nature of human involvement with ecosystems. In the second part of Figure 3.1, I have used Duncan's POET notation to represent the insistence by Hawley that for humans the relation between a population and its environment is always mediated by the organization and technology employed by that population. To Hawley this mediation not only seemed to mitigate the specificity of environment as a finite (local) territory, it appeared also to abrogate environmental limits to social progress such as were presupposed by the author of the famous 1798 essay on population pressure, Robert Malthus (and seemingly accepted by Odum).

This diagrammatic representation of Hawley's idea (using Duncan's notation), with O and T enclosed between two arcs, resembles a lens. It thus seems to reflect Hawley's view that organization and technology could magnify an environment's carrying capacity. However, as the symmetry of the diagram reveals, it is equally plausible to imagine looking through the lens from the E side, in which case O and T would magnify P. Indeed, organization and technology *have* enlarged the resource appetites and environmental impacts of various human populations.

Next, I applied the Duncan notation to the conception of human ecology put forth by Park, the Chicago sociologist credited with first using the term *human ecology* more than sixty years ago. Park (1936) differentiated human ecology from plant and animal ecology by pointing out the need "to reckon with the fact that in human society [there is a] cultural superstructure [that] imposes itself as an instrument of direction and control upon the biotic substructure." Pursuant to that difference, Park spoke of a social complex comprising three elements — population, artifact (technological culture), and customs and beliefs (nonmaterial culture). In the third part of Figure 3.1, these ideas are represented with the POET notation. As a result, the social complex is seen as an entity; interaction occurs between it and its environment, not just between each of its component variables and the environment, or between P and E mediated by O and T.

The third model of ecological reality is, I submit, superior to either of the first two described here. For ecosocial theory purposes, this representation of Park enables us to think of O and T as modifications of P (Winner, 1986), and I propose therefore to construe Park's work as recognition of some new (ecosocial) taxa, which, without waiting for official acceptance of the idea by systematists, can be referred to as *Homo colossus*. This article will demonstrate the appropriateness of this designation.

Prosthetic Polymorphism

With different organizations and technologies, one population of humans can be a very different sort of ecological entity than another human aggregate. Accordingly, let us invoke the biological concept of polymorphism. It has been defined somewhat simplistically by Topoff (1981) for a colony of social insects as "the existence of individuals that differ in both size and structure," and defined more generally, yet more precisely, by Ford (1955) as "the occurrence together in the same habitat of two or more distinct forms of a species in such proportions that the rarest of them cannot be maintained merely by recurrent mutation."

Leaving aside the issues of genetics and natural selection implicit in Ford's definition, let us consider division of labor, a classic topic in sociology that was recognized as early as 1893 to be an extension of the biological phenomenon of organic specialization (Durkheim, 1933). Division of labor arises even in very simple human societies, based at least on age and sex differences. In modern societies it becomes much more elaborate (Catton, 1985).

For the task of modeling ecosystem processes when humans are involved, we need to broaden the concept of polymorphism. I propose that the possible differentiation of functions is limited when it has to depend either on biological polymorphism within a single species or on genetically based differences between the various species cooperating in a biotic community. Human societies have transcended these limits, and, ecologically speaking, what is distinctive about our species is that we have substituted sociocultural differentiation and technology for biological polymorphism and interspecific differences.

This is the way biologists' and sociologists' views of the special nature of the human species ought to converge. Among the members of a human labor force, the polymorphism that makes possible a highly ramified division of functions is in the tools rather than in the hands that wield them. There is polymorphism in the socially instilled contents of the brains that control those hands and those tools, not in the biological structure of those brains.

Machines, tools, and other artifacts can be described as *prosthetic organs* — detachable extensions of the human body. The British Museum of Natural History in London has an eloquent exhibit on natural selection that includes a display comparing variations in an organ with variations among tools adapted for different tasks. It acquaints viewers with the ecological significance seen by Darwin in the assorted types of beaks on the several species of finch he observed exploiting different resources on the Galapagos Islands (British Museum of Natural History Staff, 1981). To advance the task of modeling ecosystem processes in which humans are involved, we should therefore broaden (in a sociological direction) our use of the term *polymorphism*. If we consider an industrial civilization's human labor force not just as a population of furless and bipedal mammals but instead as a population of social complexes (tool-users-modified-by-their-respective-tools-and-organizational-roles), then it can be seen that our species is impressively polymorphic. Culture enables *Homo colossus* to be, in this sense, the world's *most* polymorphic species.

This important ecological implication of culture has new significance for reuniting sociology and biology. Substitution of human sociocultural polymorphism for the less diversified and much less flexible biological version has important consequences. Ecologists and sociologists should stress this fact to the public, who may then require such understanding among elected policy makers.

Sociocultural polymorphism had already been recognized (without the label) when Colinvaux (1973, p. 579) wrote, "Man alone can change his niche without speciating." I prefer to speak of quasi-speciation, meaning the adaptation of various members of the one human species to different niches by cultural (i.e., technological and organizational) differentiation without recourse to genetic differentiation. It is by this means that humans have in the course of their evolution several times succeeded in usurping from other species portions of the planet's total life-supporting capacity. Each time, human numbers increased. Now it is essential to see that the latest episodes of quasi-speciation can lead to resource scarcity and environmental degradation. Colinvaux (1973, p. 579) recognized that "The time is already on us when ... the carrying capacity of our living space is not enough to provide a broadened niche for all men who now exist."

Traps

Freese (1985) provides a clear definition of a *serial trap*, further elucidating issues raised by the now famous description of the *commons dilemma* by Hardin (1968). A serial trap exists when resources required by a user population are replaced over time at a more or less constant rate; replacement rate is exceeded by use rate; resource depletion cumulatively affects further availability, so that relative scarcity intensifies exponentially; and, as time passes, system degradation becomes less and less reversible. As Freese notes, serial traps clearly do occur in natural ecosystems not under human domination. Various species in various ecosystems have experienced the cycle of irruption and crash (e.g., birds, caribou; see Remmert, 1980, Welty, 1982). But he seeks to persuade sociologists that serial traps also occur in human-dominated ecosystems (Whittaker, 1975), and that the dependence of industrial societies on nonrenewable resources must be seen as an example. (By definition, the replacement rate for *nonrenewable* resources must be effectively constant, i.e., zero, and any nonzero rate of use must exceed it.)

Modern societies have consistently mistaken rates of discovery for rates of replacement (Pratt, 1952; Simon and Kahn, 1984), entrapment being the result of the illusion that all is well if use rates are just not yet in excess of recent discovery rates.

What political and economic decision-makers and their constituents most need to learn from an ecosocial theory is the idea that the cumulative effects of ecosystem use can make it progressively less feasible to retreat from an accustomed use pattern back to an earlier one after the newer pattern belatedly comes to be seen for the trap it is (see Costanza, 1987, p. 407).

Quite recently, man–machine combinations enlarged the effective environment, but precariously so. Between 1930 and 1960, most draft animals on U.S. farms were replaced by tractors. According to the Office of Technology Assessment (1985), this released some 20% of U.S. cropland from raising feed for animals and made it available for growing crops for human consumption. Conventional wisdom accepts this as unmitigated progress. It is not seen as a trap. Ecologists may astutely ask, however, what is to happen after humans have expanded their numbers or their appetites in response to the 20% capacity increment? If the fossil fuels for tractors become depleted and too costly, some land may again need to be devoted to producing biomass fuel (either for the tractors or as feed for a new generation of draft animals).

The OTA (1985, p. 19) went on to say, "The increased mechanization of farming permitted the amount of land cultivated per farm worker to increase fivefold from 1930 to 1980." For purposes of ecological modeling, it is as if farmworkers (as PTO complexes, not just as P) had been enlarged by a factor of five; each can do five times as much farming as could his less colossal grandfather. How was this enlargement accomplished?

According to the OTA (1985, p. 19) report, "The amount of capital ... used per worker increased more than 15 times in this period," and furthermore there is now heavy reliance "on the nonfarm sector for machinery, fuel, fertilizer, and other chemicals." Clearly, then, the farm labor force is not just P; O and T can sensibly be viewed as extensions of it. But again, we need to recognize as a trap this conversion of *Homo sapiens* into *Homo colossus*.

Carrying Capacity and von Liebig's Law

The concept of carrying capacity, if correctly understood, can spotlight traps. For any use of any environment by any population, there is a volume and intensity of use that can be exceeded only by degrading that environment's future suitability for that use. Carrying capacity, the word for maximal sustainable use level, *can* be exceeded — but only temporarily. Ecologically, Malthus's main error was supposing that it was not possible for a population to increase beyond the level of available sustenance. It can and does happen, but the overshoot will always be temporary.

The comparably tragic error of Malthus's latter-day critics has been to mistake serial traps for progress, i.e., to construe technological change that facilitates temporary evasion of carrying capacity limits as permanent elevation (or repeal) of those limits. When load comes to exceed carrying capacity, the overload inexorably causes environmental damage; then the reduced carrying capacity leads to load reduction (i.e., a crash).

Ecologists have not made this situation clear enough. Too often they have embraced the logistic curve model for population growth and have construed the upper asymptote as the best representation of carrying capacity (e.g., Emlen, 1984). The logistic curve does not rise above that upper limit, and the limit is represented by a constant in the mathematical formula. But carrying capacities are not constant; they can and do change.

Political and economic leaders, and social scientists for that matter, tend to exaggerate any recognition that carrying capacity is not constant into the supposition that it is infinite. The fact that carrying capacities can be difficult to measure cannot exempt populations from the consequences of exceeding their environment's power to sustain them. The human prospect would be brighter if somehow these points were to be central to the agenda for the next superpower summit meeting — but, of course, they won't even be mentioned.

Recently there has appeared both a biological and a sociological literature alleging an inescapably subjective element in so-called carrying capacity. It is said that carrying capacity ultimately depends on people's value judgments (McHale and McHale, 1976; Shelby and Heberlein, 1984; Wagar, 1964). Some imply that, unless carrying capacity can be assigned a precise value, the concept has no significance. Politicians and industrialists grasp at such straws all too eagerly.

The antidote to such thinking is provided by the relation between the carrying capacity concept and von Liebig's law of the minimum. Justus von Liebig, an agricultural chemist, showed (1842) that it was the least abundantly available nutrient that limited the yield a farm could produce. It need not be difficult to see why this must be so (and why it can be generalized to so many phenomena), given that any organism is a complex chemical structure. The number of specimens of any organism that can be constructed from a given assortment of chemical components will be limited by the scarcest component. (The principle also can be illustrated with a nonbiologi-

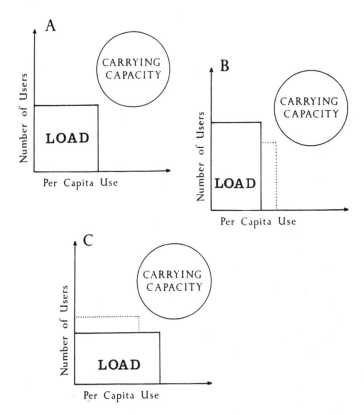

Figure 3.2   Sustainable load: three versions. Load is the product of two dimensions: the number of users and the mean per-capacity use. A sustainable load must not exceed the sustained rate of supply, the carrying capacity. An increase along one axis of the graph must be compensated by a decrease in the other dimension. All three graphs in this figure represent sustainable loads.

cal example. Imagine a collection of a dozen flashlight batteries, five battery cases, six reflector and lens assemblies, and two three-volt bulbs. The availability of only two bulbs will limit to just two the number of working two-cell flashlights that can be assembled, even though there will be other parts left.)

With this in mind, let us see why it is so misleading to imply or assert that the concept of carrying capacity is based on value judgment. In Figure 3.2A, I have represented carrying capacity by a circle and load by a square (drawn so that its area is equal to the area of the circle). Think of the circle as the cross-section of a pipeline through which there is a constant flow of some limiting resource. Quantitatively, an environment's carrying capacity for a particular life form is set (according to von Liebig's law) by the continual rate of flow of the least abundantly available necessary resource. The load is clearly the product of two dimensions: the number of users of that limiting resource multiplied by the mean per-capita rate of use. The point is this: a sustainable load is a load not exceeding the sustained rate of supply.

Clearly, the load may have different shapes and still be compatible with carrying capacity. Instead of the schematic square, we may substitute a vertical rectangle (Figure 3.2B), representing an increase in the number of users, and a commensurate reduction in per-capita mean use of the limiting resource. As long as the area of the rectangle remains no larger than the area of the circle, we have a representation of a sustainable load. Alternatively, we could have a horizontal rectangle (Figure 3.2C) where per-capita use level has increased and the tradeoff enabling the load to remain sustainable is a reduction in user numbers. From these comparisons what is most vital to note is that the question of a load's sustainable magnitude is an objective ecological issue, not a value question. The question of which tradeoff is preferable to its alternative is a value issue. Choosing whether to increase the user population at the cost of lowering its standard of living or to raise affluence at the cost of population reduction depends on a value judgment. But it is a serious mistake to suppose that this denudes carrying capacity of any objective meaning.

The importance of avoiding such erroneous thinking becomes clear when we imagine increasing one dimension of load without the tradeoff on the other dimension (Figure 3.3). If population growth continues so that a maximum sustainable load has an overload added to it, habitat damage takes a bite out of carrying capacity. Likewise, if per-capita use rises beyond the level prevailing in an already maximal load, and there is no tradeoff reduction of user numbers, the overload must again result in habitat damage and carrying capacity reduction.

The normative questions that have led some analysts to declare carrying capacity a useless concept have to do with questions of equity rather than of sustainability. Neither biologists nor sociologists should confuse the two. Political disagreements as to what constitutes equitable allocation of finite resources should not obscure the fact that nature exacts penalties when loads exceed carrying capacity, whether the excess comes on the vertical or the horizontal dimension.

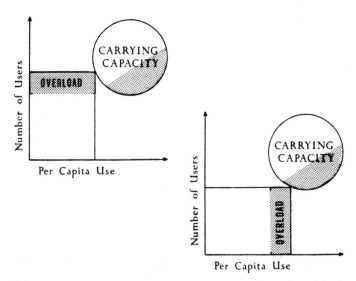

Figure 3.3 Ecosystem-damaging overloads. If one dimension of the load increases without compensatory decreases, the overload diminishes the carrying capacity. Loss of carrying capacity is represented by the shaded area of the circle.

Actually, the human load has been expanding on both dimensions. The number of humans on earth has increased enormously since prehistoric times. There has also been great technological progress over the millennia, especially in the last two centuries. We are not yet accustomed, however, to putting these two items of knowledge together and recognizing the two-dimensional enlargement (or the enormity of that enlargement) of the human load, nor have we come to terms with its ecological implications.

Homo Colossus

This two-way expansion of the human load is represented graphically in Figure 3.4. For many warm-blooded species, to maintain life with no gain or loss of body substance an animal needs an average daily food energy intake of

$$w^{3/4} \times 70 \text{ kcal,}$$

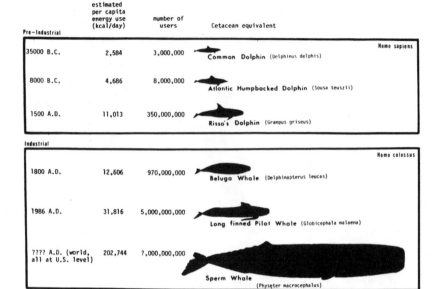

Figure 3.4 The changing human load, as represented in cetacean equivalents. Over the last 37,000 years, not only has the number of humans increased, but also the amount of energy each consumes. At one time, each person used approximately the same amount of energy as a common dolphin. Today in the United States the average person uses as much energy as a sperm whale. How many such *Homo colossus* individuals can the earth support?

where w is body weight in kilograms (Kleiber, 1947). Applying this formula to size estimates for various cetacean species (Gaskin, 1982, Minasian et al., 1984) and to estimates of human exosomatic energy use plus food intake (Catton, 1986) enables us to select particular dolphin or whale types to represent various stages in the evolution of *Homo sapiens* into *Homo colossus*. (It is cultural evolution, not biological evolution, we thus represent.)

The human load to be supported by the ecosystems of the world has not just grown from 3×10^6 individuals in 35,000 B.C. to 5×10^9 equivalent individuals today. Each of those three million hunter-gatherers was the energy-using counterpart of a common dolphin (*Delphinus delphis*), whereas each of today's 232 million Americans matches the energy use of a sperm whale (*Physeter macrocephalus*).

Projecting that all humans can someday be as industrialized as Americans have become (Kahn et al., 1976) is equivalent to imagining a world populated by five billion sperm whales. It reflects woeful ignorance of the ecological consequences of cultural polymorphism. Such is the folly implicit in declaring that "the term *carrying capacity* has by now no useful meaning" (Simon and Kahn, 1984, p. 45).

We urgently need widespread dissemination of the fact that carrying capacity is not infinitely expandable. The time has come when "tragic choices" must be acknowledged (Calabresi and Bobbitt, 1978). In the world as it is and is going to be, human loads can grow on one axis only by shrinking on the other axis. Otherwise, our legacy to posterity will be reduced carrying capacity and the human suffering that it will entail. Anyone wishing for a more humane and happier future should strive to spread ecological literacy. Those who aspire to leadership should be required to demonstrate that they understand the load and carrying capacity concepts.

How to Prosper in a World of Limited Resources: Lessons from Coral Reefs and Forests on Poor Soils

Eugene P. Odum

J believe it was Will Rogers who once remarked that the Good Lord continues to make more people, but he isn't making any more land. The world population of humans will probably double at least one more time (from 5 to 10 billion) before, hopefully, leveling off sometime in the next century. In any event, space and many nonreplaceable resources (i.e., resources having no conceivable substitutes such as clean air and water) will increasingly be impacted or in short supply. Whether we like it or not, society will have to learn how to do with less. We may be able to learn how to do this by studying natural ecosystems that are adapted to thrive on scarce resources.

The Coral Reef

No one should go through life without at least once donning a face mask and snorkel and taking an underwater look at a coral reef. What you see is a fantastic natural city. Brightly colored fishes rush around in and over the limestone structures that have been built by the coral animals and the coraline algal plants. The structures vary from massive formations such as "brain

---

Abstracted from 1990 Phinizy Lecture, University of Georgia.

corals" to tall, branched "skyscrapers," often brightly colored. There are channels paved with coral sand that resemble streets that wind around and between the coral "heads," as the formations are called. Moving along these channels, you may notice a fierce-looking moray eel lurking in a crevice or limestone cave; he won't bother you if you don't bother him. Or you may meet a small shark; he won't "mug" you as long as you give him a chance to escape. Bigger sharks that might be a threat will be in the deeper waters adjacent to the reef, where one is advised to swim with caution. You may encounter piles of rubble where coral structures have been knocked down by storms, or perhaps just collapsed with age. As in the human city, the reef is constantly breaking down and being repaired.

The coral reef has many inhabitants, such as lobster, shrimp, and octopus, that are most active at night when the coral animals extend their tentacles from the thousands of little pockets where they live. The corals do this in order to capture microscopic "plankton" (free floating life) suspended in the water. At night the coral formations appear to be covered with tiny feathers waving in the current. It reminds one of people hanging out of windows waving flags during a parade. Thus, a visit to the reef at night reveals a "nightlife" that is different from what one sees in the daytime. But you have to bring a strong underwater flashlight since there are no streetlights.

What is truly remarkable about coral reefs, especially those associated with Pacific atolls, is that they are able to support such concentrated and diverse life even when surrounded by nutrient-poor waters; the reef is like a prosperous oasis in the middle of a desert. Unlike human-made cities, there are no huge truckloads and trainloads of concentrated food, building materials, and other resources coming in every day from somewhere else. However, there is plenty of sunlight and adequate calcium in the water for building the basic reef structure. The trade winds and tides provide the energy for water flow that is the transportation system for the reef. But many vital nutrients such as nitrogen and phosphorus that are the building blocks of all life are extremely scarce. How, then, does the coal reef prosper in its world of very limited material resources?

Some 35 years ago my brother and I became interested in this question. We obtained a government grant to carry out a study of the metabolism of a coral reef on Eniwetok Atoll in the Marshall Islands. For our study we selected an inter-island reef where the water flow is one-way rather than back and forth as is the case for reefs that border the islands. With water flowing over the reef in one direction (from sea to lagoon), we could measure the quantity of nutrients coming in and going out. Also, we could estimate food production and consumption for the whole reef by comparing daytime and nighttime changes in oxygen in the water as it passes over the reef. It is during the day that oxygen is being produced by photosynthesis, while at night oxygen is only being consumed by the respiration of all of inhabitants. With these sorts of data we were able to construct a balance sheet of inputs and outputs and estimate productivity, in much the same way one might assess the performance of a factory or a business.

It was evident that the reef was extremely productive. Our calculations suggested that the rate of production of organic matter (i.e., "food" that supports all the incredible variety of organisms) on a unit area and time basis (grams or calories per square meter per day or year) was as great as that of a heavily fertilized Iowa cornfield. Yet there was very little fertilizer coming in from outside; the phosphorus in the incoming water, for example, was so low that we were barely able to measure it with the equipment available. The logical explanation, of course, is that nutrients are being retained and efficiently recycled within the reef so that very little is needed in the way of imports.

We also found that, despite the fact that the coral polyp (the technical name for the individual animal in the colony) is well adapted to capture whatever plankton is available, there was not nearly enough of it coming across the reef to support the very large populations of coral animals. We suggested, but were not able to demonstrate at that time, that corals were obtaining food directly from the algal cells that grow in and around the animal tissues. Living corals, by the way, unlike the bleached white skeletons in the curio shops, are greenish, although some species have yellow, blue, or other pigments that hide the green. Later, several investigators were able to show by using tracers that there is indeed an active and direct exchange of food and nutrients between coral animals, which excrete nutrients such as nitrogen and phosphorus, and algal plants, which are capable of using sunlight, carbon dioxide, and mineral nutrients to make food.

The idea that the coral is an interdependent plant–animal cooperating complex or system was a radical notion at the time, but it is now accepted as one of the main reasons why coral reefs prosper in a nutrient-poor environment. In other words, the coral gets its "vegetables" from algal gardens it cultivates and fertilizes within its limestone house and gets its "meat" from hunting and gathering in the adjacent environment. Recent studies indicate that no more than 10–20% of the polyp's daily caloric requirement comes from animal plankton (zooplankton), but this plankton is important as a supply of protein.

The coral is not the only animal on the reef that has algae in its tissues. The "lips" (mantle) of the giant clam, for example, are bright green; the clam opens its shell not only to filter out microscopic life from the water but to expose its plant tissues to the sun. A recent study has shown that as much as 80% of the total food (i.e., photosynthate) manufactured by all the algae on the reef that is not used to support the algae themselves is directly transferred to the animal components of the reef.

We can now say that *close encounters of the mutual kind* is the basis for the coral reef's success in overcoming resource limitation. Producers and consumers (plants and animals) live in intimate contact and aid each other. There is little waste that is not used or recycled by some member of the community. Sunlight and water flow are efficiently used as energy sources. The diversity of species insures that all the "jobs," or ecological niches, necessary for an efficient and regenerative "economy" will be carried out by some critter or group of critters.

The lesson we learn can be stated as follows: *When things get tight and resources are scarce, it pays to cooperate for mutual benefit.* Or, we might even say, *we have to work together in order to survive.*

It is a fantastic idea, but could we possibly design a future human city organized on the coral reef plan where everything possible is recycled, every inhabitant grows at least some of their vegetables in a greenhouse room (using artificial light where sunlight was not conveniently available), and goes to market mostly for meat and specialty items? With the human world becoming increasingly urbanized (one prediction: 80% people will live in cities in the next century), something of this sort may be necessary to help feed the cities and to lift the terrible burden of waste and pollution that current cities now cause.

Other Examples of Adaptations to Limited Resources

The coral reef is by no means the only large-scale adaptation to scarcity to be found in nature. Where soil is poor, water scarce, or temperatures extreme, we find that natural communities have found ways not only to survive but even to prosper with whatever resources are available. Two other examples will suffice to illustrate.

Pines to the Rescue

During the early part of this century, much of the six inches or so of good topsoil that nature had built on the Piedmont, as the hill region of southeastern United States is called, was eroded away as a result of very bad farming. Small farmers, many of them tenants or "sharecroppers," had been forced by the social and economic conditions of the time to plow up the hills in order to plant cotton and other summer "cash crops" required to pay their bills and their rent, in many cases to absentee landlords. There were no incentives to plant protective winter and spring "cover crops" because they didn't have an immediate market value. Accordingly, the plowed ground was exposed to heavy rains throughout the year. By the 1930s, much of the Piedmont landscape was eroded down to the red subsoil, and the region had become poverty-stricken.

Fortunately, nature and the federal Soil Conservation Service came to the rescue. Not only were terracing and cover crops promoted, but southern pine trees, which are able to grow on soil so poor that you couldn't grow corn or tomatoes on it, invaded the abandoned and eroded fields or were planted on them. Today, the hills are covered with reasonably prosperous forests that are gradually building back the soil.

Like the coral, the success of the pine in coping with nutrient scarcity involves a "close encounter of the mutual aid kind." Pine trees have developed a partnership with special kinds of fungi called mycorrhizae, or "root fungi" ("myco" is latin for fungus, "rhiza" for root). The root fungi greatly expands the root system. Furthermore, the fungi have a special talent for extracting phosphorus and other nutrients from poor soils. The slender fungal threads cost less energy to produce and maintain than roots, and they can grow out into the soil far beyond where roots can go. In return for the valuable nutrient-collecting service, the pine provides the fungi with photosynthate food that the fungus, being a non-green plant, is unable to produce for itself.

Pines inoculated with a very heavy crop of mycorrhizae are just about the only trees that are able to survive and grow on very severely eroded land devastated by acid fumes from copper smelters in Copper Hill, Tennessee. In this case, sewage sludge was used to give the pines a head start in the seemingly impossible job of colonizing this man-made desert. Once the pines and their "mutual aid" system are well established, the forest can make it on its own without additional fertilizer.

The Tropical Rain Forest

A tropical rain forest on sandy infertile soils of the Amazon Basin is an even more spectacular example of prospering despite limited resources. Mycorrhizae play an important role here, but so do a host of other adaptations that contribute to retention and efficient recycling. For example, rain forest leaves are thick and waxy, which retards loss of nutrients in the heavy rains and also discourages leaf-eating insects. Older leaves become covered with nitrogen-fixing lichens that can convert nitrogen in the air, which is not usable by trees, to nitrate, which is. When leaves do fall, fine roots penetrate the leaf litter and quickly recover nutrients in fallen leaves before they are "leached," that is, lost in the porous soil. Some rain forest trees even have "upwardly mobile roots" that grow upward in the bark furrows at the base of the tree (instead of downward as do normal roots) and thus are able to recapture nutrients from rainwater flowing down the trunk. So the whole forest is one remark-able recycling system!

On the Iowa prairie the fertility is in the deep, dark soil where the mineral nutrient "goodies" are stored. Replacing the native prairie grasses with grain (i.e., cultivated grasses) is not a radical change that reduces the potential productivity, provided, of course, that soil erosion is controlled. In striking contrast, *the fertility of many rain forests is not in the soil but in the living structures* ("biomass" in ecological language), *where almost all of the "good-ies" are stored and recycled.* Removing the forest thus completely destroys the adapted mechanisms and reduces the productivity manyfold, as is all too evident when one tries to grow crops on the deforested land. In the tropics, large trees and verdant vegetation do not necessarily indicate fertile soil.

Before the advent of population explosions, urbanization, and invasion by wealth-seeking Europeans, native forest people fed themselves by a system of "shifting agriculture." They cleared and burned small plots and raised crops for two or three years until all the residual nutrients were exhausted. Then the farmer moved on to another place, leaving the forest to reestablish and restore fertility; it could do that fairly quickly since all the variety of species and conditions were available close at hand. Tropical peoples includ-ing the prosperous Mayans of Central America also developed sophisticated "mutual aid" combinations of agriculture and horticulture involving mix-tures of annual, perennial, and tree crops, domestic animals, and fish ponds fed with manure and plant residues. These kinds of "traditional agriculture" are capable of feeding villages and small cities on a sustainable basis but do not produce a lot of surplus to feed large cities.

Today, radical changes are coming to the tropics. Forests are being cleared on such a large scale as to threaten the stability of world climates, and traditional agriculture is being replaced by attempts to impose a northern-style monoculture (one species) cropping. Both these trends represent a desperate attempt to find a place for thousands of homeless and to supply food and monetary wealth for cities that are growing too fast for their own good. Scientists and thoughtful people everywhere are searching for alternatives that might avoid what would seem to be impending ecological and economic disaster.

What Can We Learn From Nature's Ability to Get More from Less?

The coral reef, the pine tree, and the rain forest give us clues to how to "power down" (i.e., increase efficiency and reduce waste) and yet remain prosperous. And these examples teach us that cooperation, and "mutual aid," are the keys to success when things get tough.

We can begin to heed this "wisdom of nature" by greater use of already existing technology that will enable the industrial nations (and the rich in general) to power down gracefully without loss of reasonably good living standards, and at the same time establishing mutual aid networks with less-developed countries (and the poor in general) to enable them to power up to achieve a decent standard of living. For example, the technology is already here to reduce by half or more the amount of fuel needed to travel a given distance. New farming techniques involving "conservation tillage" and "crop rotations" are already demonstrating that the crop yields we now get can be sustained with half the pesticides and fertilizer we now use. Less-developed countries can go directly from their traditional agriculture to these new technologies in order to bypass the current "high-chemical-input" agriculture that is so damaging to the environment. The technology of removing acids and other contaminants from coal before it is burned in power plants so as to reduce air pollution has been known for years now (but not yet applied on any large scale). And, of course, the techniques of birth control and family planning are improving every year.

What is lacking is, first, a worldwide public awareness of the critical state that our big house, the earth, is in; second, demonstration that changes that can take the earth off the critical list are economically feasible (in fact, reducing wastes and costs can increase profits); and, third, the will (both at the individual and government levels) to undertake some fundamental changes to ensure that we can prosper (not just survive) in a more crowded and demanding human world with fewer unused and easily exploitable resources.

How Universities Should Grow

Eugene P. Odum

*I*f you look, as ecologists do, at the way things grow, you'll see some interesting parallels. Whether you're examining ecosystems, cities, societies, academic disciplines, or universities, the patterns of growth and change are remarkably similar.

In ecology, we often refer to the "sigmoid growth form," a term used to describe a three-phased growth period in which an interval of slow growth is followed by very rapid, often exponential, growth in size and complexity, and eventually levels off to a kind of pulsing plateau.

This is the "S-curve" for quantitative growth. Theodore Modis, a physicist, notes the widespread occurrence of this pattern in a recent book review he wrote for the journal *Science* (**259**:1349, 1993). His review of the book *Predictions* carries the portentous title "S-Curves Everywhere."

Sometimes the rapid growth phase continues for too long, producing a "boom-and-bust cycle." This happens in nature, and too often in human affairs as, for example, when corporations grow too fast and become too big for their own good, and then have to scale back to survive. The recent history of IBM may be an example.

My theme in this commentary is that there is another, more important — and more difficult to measure and quantify — phase of growth that can occur after growth has reached a plateau. For want of a better term, we can call this

---

Reprinted from the ISEE Newsletter, January 1995, p. 3.

phase "qualitative growth," that is, getting better rather than bigger. In other words, there is potentially a better life after the S-curve growth is complete.

We know this is true from personal experience. When growth in body size stops after adolescence, we devote most of our lives to becoming better — not bigger — human beings.

Herman Daly and his colleagues at the World Bank, in a 1991 UNESCO report titled "Environmentally Sustainable Economic Development," echo the call of a new generation of economists who argue that the time has come for a transition from traditional economic growth to qualitative growth that they call "economic development." They suggest that this transition should begin in the developed world, where waste, pollution, and per-capita consumption of resources are so unnecessarily large as to threaten global life-support systems. So far as anyone knows, there is no limit to growth in quality, but it is becoming increasingly evident that there are limits to quantitative economic growth that is based on ever more consumption of resources in a finite world.

During my 53 years with the University of Georgia, the university has gone through a period of sigmoid growth. Growth in size was very slow for more than a century. In 1940, when I came here to teach, the student body was about 3000, and the faculty was very small and burdened by backbreaking teaching loads and almost no money for basic research. Between 1950 and 1980, there was a period of very rapid growth in the student body, in graduate study, and in funds for research. Although enrollment of graduate students is still growing, efforts are now being made to plateau the size of the total student body and faculty by raising the standards for admission and for faculty advancement and recruitment — i.e., quality control. This shift from quantity to quality is the proper way for a university of continue to grow in the coming decades. Unfortunately, the level of state funding is currently based more on quantity than on quality; the more "warm bodies," the greater the appropriation. So there are very strong political incentives to grow ever bigger in the hopes of getting more money.

We need to develop and make more use of indices of quality in determining the level of funding, such as proportion of students in the honors program or who gets good jobs upon graduation, faculty research output, national rank of the institution and its programs, quality of teaching, and so on.

Growth in quality has its costs and benefits, as does growth in numbers. An often unrecognized common denominator in the growth of any large and complex system, human or natural, is that the cost of maintenance tends to increase as some kind of power function of the increase in size. That is, as a university (or any other system) doubles in size, the cost of maintaining order in the system (or as we ecologists put it, the cost of "pumping out the disorder" inherent in any complex system) more than doubles.

If you don't believe that, take a look at the growth of the budget for plant operations including the increasing costs of repairs, security, parking, buses, control of outdoor and indoor (i.e., "sick" buildings) pollution, and on and on. There are increasing returns of scale, as economists like to point out, early in a quantitative growth cycle, but sooner or later there are decreasing returns of scale. Have we reached this point at the University of Georgia where maintenance costs are seriously reducing funds available for increasing the quality of education and research?

The rise in maintenance costs is especially acute in the growth of research programs, especially in fast-moving areas. Most every laboratory has expensive instruments that are either obsolete or broken and awaiting repairs. As research programs grow, increasing amounts of grant money are diverted as "overhead" to help pay for space, facilities, and maintenance, leaving less for the actual research. Keeping costs and bureaucracy from getting out of hand is a tough challenge for administrators of large research programs.

A really good research university needs two things: endowments sufficient to support and keep talented research professors, and the means to support good technicians and postdoctoral students who do much of the day-to-day research and who provide vital link between the graduate student and the busy senior investigator.

What is not generally understood by the public is that the reputation of a university outside the state is based on research and publication, while the reputation within the state is based on the quality of teaching and service. These functions can be mutualistic rather than competitive if postdocs and tenured professors actively take part in undergraduate education.

During the rapid growth period in our university, there was a proliferation of departments, and, as we now recognize, too great a fragmentation of disciplines accompanied by a decline in the quality of general education. Fortunately, this trend is being reversed by efforts to combine closely related disciplines into one administrative unit, and especially by the development of cross-disciplinary institutes and centers that can deal with the large scale of real-world problems and thereby attract more outside money from the federal government, foundations, and industry. And all of this is being accompanied by coordinated efforts to increase the quality of undergraduate teaching.

Our Institute of Ecology is an example of a center that contributes to quality growth. We bring together people who are already here from a variety of disciplines for research and education that go beyond that of any one discipline. We strive to balance competition, which is a way of life in human affairs as in nature, with cooperation and symbiosis, which becomes important when things get crowded and complicated. Best of all, we promote the interfaces between disciplines, such as between economics and ecology or agriculture and biology.

In the near future we hope to focus on the most important of all interfaces — those between natural science, social science, and the humanities — where the real-world problems are most acute. We also promote off-campus satellite programs on the coast, at the Savannah River Site and elsewhere in the region that bring in additional staff and large amounts of outside money. Now that the institute has become administratively a "school," more of our research staff will be involved in undergraduate teaching and public service.

In 1967, renowned author and social critic Lewis Mumford published a commentary titled "Quality in Control of Quantity" (In *Natural Resources: Quality and Quantity*, eds. Ciriacy-Wantrup and Pearson, pp. 7–18, Univ. of California Press, Berkeley). He wrote,

> The problem of our age, which dominates and underlies most of its other problems, is the problem of quantity. I intend no diatribe against progress in science and technics, so long as they remain subordinate to organic functions and human purposes. I am merely trying to explain how their immense human benefits were curtailed by a one-sided emphasis on quantity, and the exercise of a one-sided control over both man and nature, who speedily become victim of his own favored method.

"Quality in control of quantity," Mumford continued, "is the great lesson of biological evolution." It is also a great lesson for research universities.

Energy:
The Common
Denominator

Eugene P. Odum

T he source, amount, and quality of available energy determines the kinds and numbers of organisms in nature and, ultimately, the number and lifestyle of humans. Recognition that energy is not only the common denominator of humans and environment but also the ultimate limiting factor for development is of key importance if we are to maintain human quality of life in the future. To document this overall statement, we need to review how energy is measured, how much is needed for basic life-supporting activities, the thermodynamic laws that determine how energy is used, and the importance in considering quality as well as quantity.

Units used to measure the quantity of energy unfortunately differ according to the form (electric, gas, coal, food, and so on) and the measurement system (metric or English). While "conversion factors" can be used to compare any unit with any other one, there is yet no agreed-upon international unit of energy quantity. Some energy units, such as the watt used to measure power, have time built into the definition and are thus energy–time or power units. Other units such as the calorie represent potential energy (not time-specific), and a unit of time must be added to convert these units to

Originally published as part of a *Bulletin on Energy* issued by the University of Georgia's Institute of Ecology on Earth Day in 1970.

power rates. In ecology we mostly use kilocalories (abbreviated kcal) per day or per year to quantify energy flow. In order to compare various kinds of ecosystems, we add a unit of area such as the square meter, acre, hectare, etc. To relate quantities to your own personal experience, the following might be remembered:

1. An adult in the United States consumes about 2800 kilocalories of food daily, or about 1,000,000 kcal yearly. A gram of carbohydrate or protein yields 4–5 kcal, and fat yields about 9 kcal when burned in the body.

2. There are 10,000 square meters in a hectare; 4,046 square meters in an acre.

3. Sunlight reaching the earth's surface in the temperate latitudes is about 1.0 to 1.5 million kcal per square meter per year, with 1 to 5% converted to organic matter by plant cover (photosynthesis).

4. The potential energy in a pound of coal is about 3,150 kcal, a pound of gasoline 4,230 kcal, and a gallon of gasoline 32,000 kcal.

It is extremely important to note that the quality of energy varies widely with the kind of source. For example, sunlight is abundant but low in quality in terms of potential for useful work as compared with oil, which has a much higher quality in this regard. Since there are no units yet agreed upon to measure quality, the best we can do is to use ratios. It can be estimated that one kcal of oil is equal in work capacity to about 2,000 kcal of sunlight; the quality of oil energy is therefore 2,000 times that of solar energy. Thus, switching man's cities from fuel to sun power involves dealing with a much larger quantity of the latter and requires the development of new and expressive concentrating processes.

Energy Laws. The thing to remember about energy is that, unlike materials, energy cannot be reused or recycled. When energy is converted from one form to another, a large part is dissipated into unusable heat, so there has to be a continuous flow of energy to maintain any system of humans or nature. The good news is that in a series of transformations the quality may be increased. Thus, when sun energy is transformed into food by plants, or into electricity by photovoltaic panels, the quantity goes down but the quality goes up.

Kinds of Energy Systems

The per-capita fuel energy consumption in the United States approaches 100 million kcal per year, while such energy consumption per square meter of a city approaches one million kcal annually. Whether man-made or natural, ecosystems rely on two major sources of energy: the sun and chemical or nuclear fuels. We can conveniently distinguish between solar- and fuel-powered systems on the basis of major input while recognizing that in any given situation both sources may be utilized. It is important to note that, although the total solar energy impinging upon the earth is enormous, solar radiation on an area basis is a diluted energy source. Only a small portion of the solar radiation that falls on a square meter is directly usable by organisms (as noted above). The power density of solar radiation can be raised considerably — even by a factor of ten — through auxiliary sources of energy (such as wind, tides, and fuel). These energy subsidies reduce the cost of self-maintenance of the ecosystem, and thereby increase the amount of solar energy that can be converted into organic production. In contrast to the solar power systems, the fuel that supports the fuel-powered systems provide a highly concentrated source in terms of conversion of useful work within a small area. These solar- and fuel-powered ecosystems can be divided into four categories:

1. Unsubsidized natural solar-powered ecosystems (forests, grasslands, etc.).

2. Naturally subsidized solar-powered ecosystems (example, tidal estuaries).

3. Man-subsidized solar-powered ecosystems (as in agriculture).

4. Fuel-powered urban-industrial ecosystems.

Unsubsidized Natural Solar-Powered

The systems of nature that depend largely or entirely on the direct rays of the sun can be designated as unsubsidized solar-powered ecosystems. They are unsubsidized in the sense that there is little, if any, available auxiliary source of energy to enhance or to supplement solar radiation. The open oceans, large tracts of upland forests and grasslands, and the large deep lakes are examples of relatively unsubsidized solar-powered ecosystems. Frequently, they are subjected to other limitations as well, as, for example, a shortage of nutrients or of water. Consequently, ecosystems in this broad category vary widely, but are generally low-powered and have a low productivity, or capacity to do work. Organisms that populate such systems have evolved remarkable adaptations efficiently using scarce energy and other resources.

While the "power density" of natural ecosystems in this first category is not very impressive, nor could such ecosystems by themselves support a high density of people, they are nonetheless extremely important because of their huge size. The oceans alone cover almost 70% of the globe. From the human interest standpoint, the aggregate of solar-powered natural ecosystems can be thought of, and they certainly should be highly valued as, the basic life-support module for Spaceship Earth. It is here that large volumes of air are purified daily, water recycled, climates controlled, weather moderated, and much other useful work accomplished. A portion of man's food and fiber needs are also produced as a byproduct without economic cost or management effort by man. This evaluation, of course, does not include the priceless aesthetic values inherent in a sweeping view of the ocean, or the grandeur of an unmanaged forest, or the cultural desirability of green open space.

Naturally Subsidized Solar-Powered

A coastal estuary is a good example of a natural ecosystem subsidized by the energy of tides, waves, and currents. Since the back-and-forth flow of water does part of the necessary work of recycling and transfer of food from one organism to another, the organisms in an estuary can concentrate their efforts on more efficient conversion of sun energy to organic matter. In a very real sense, organisms in an estuary are adapted to utilize tidal power. Consequently, estuaries tend to be more fertile than, say, an adjacent land area or a pond that receives the same solar input, but does not have the benefit of the tidal and other water flow energy subsidies. Subsidies that enhance productivity can take other forms, as, for example, wind and rain in a tropical rain forest, the flowing water of a stream, or imported organic matter and nutrients received by a small lake from its watershed.

Man-Subsidized Solar-Powered

Man, of course, learned early to modify and subsidize nature for his direct benefit, and he has become increasingly skillful in not only raising productivity, but more especially in channeling that productivity into food and fiber that are easily harvested, processed, and used. Agriculture (land culture) and aquaculture (water culture) are the prime examples of the man-subsidized solar-powered systems. High yields of food are maintained by large inputs of fuel (and in more primitive agriculture, human and animal labor) in cultivation, irrigation, fertilization, genetic selection, and pest control. Thus, tractor fuel, as well as animal or human labor, is just as much an energy input in agroecosystems as is sunlight. It can be measured as calories or horsepower expended not only in the field, but also in processing and transporting food to the super market. As another ecologist, H.T. Odum, has so aptly described the system, the bread, rice, corn, and potatoes that feed the masses of people are partly made of oil.

Fuel-Powered Urban-Industrial

We now come to man's crowning achievement, the fuel-powered ecosystem, otherwise known as the urban-industrial system. Here, highly concentrated potential energy of fuel replaces rather than merely supplements sun energy. As cities are now managed, solar energy is not only not used, it becomes a costly nuisance by heating up concrete, contributing to the generation of smog, and so on. As fuel becomes more expensive for man, it is likely that interest in utilizing solar energy in cities will increase, so we can anticipate a new class of ecosystems, the "sun-subsidized, fuel-powered city." Also, man may find it prudent to develop a whole new technology designed to concentrate solar energy to a level where it might partially replace fuel rather than merely supplement it. Only time will tell what should be man's best strategy for survival, but it will have to be based on a better partnership between man and nature than what exists now.

As of now, two properties of the fuel-powered system need to be emphasized. First, we should take note of the enormous energy requirement of a densely populated urban-industrial area. The kilocalories of energy that annually flow through a square meter of an industrialized city are to be measured in the millions rather than the thousands. Thus, in an acre of highly developed fuel-powered urban environment where most U.S. citizens live, a person consumes approximately 87 million kcal per year. Recall from an earlier paragraph that only 1 million kcal per person is required for food energy. Thus, household, industrial, commercial, transportation, and other "cultural" activities in the United States use 86 times as much energy as that required for man's physiological needs such as food energy to power the body. In underdeveloped countries, of course, the situation may be quite different. Per-capita fuel energy consumption in India and Pakistan is 1/50 to 1/100 as great, respectively, as the energy consumed in the United States. In such countries, human and animal labor are still more important than machines, and a much larger proportion of the country's total energy flow involves food and food production.

During the past decade or so, per-capita energy consumption has been increasing at a much faster rate than population growth. By the time you read this booklet, annual per-capita consumption in the United States will probably be well past 90 million kcal. Such a disparity is a matter of grave concern. For one thing, the rich tend to get richer faster than the poor under such an unbalanced growth pattern, which could lead to social upheavals that could bring on wars of destruction.

The second point to emphasize is that the fuel-powered system, in contrast to a naturally powered one, is an incomplete or dependent ecosystem in terms of life support. The fuel-powered system produces no food, assimilates very few wastes, and recycles only a small portion of its water and other material needs. Thus, an acre of a city requires not only many acres of agroecosystems to feed it, but even more acres of general life support, natural, or semi-natural environment to take care of the carbon dioxide and other large-volume wastes, and to supply it with high volumes of water and other materials. The per-capita use of water, including irrigation, is around 2000 gallons per day, of which 730 gallons are consumed (that is, not returned to streams or to other resources). A city person also consumes a ton of wood products (paper and lumber) a year, which requires from 0.3–1 acres to produce, depending on the intensity of forest management. These are just two examples of an affluent individual's impact on his environment.

To summarize, the stress that a high-powered fuel system places on the adjacent lower-powered sun system is enormous and proportional to the power differential between them, which of course increases with the power level of the city since there is a sharp upper limit to the work capacity of any system powered by dilute sun energy. The richer the city in terms of energy use, the greater the area of life support that is required. City planners and developers are often strangely unaware of this reality. It is no accident that all of the world's great industrial cities are located on coasts, large estuaries, large rivers, or fertile deltas where the life-support capacity of the natural environment is high or extensive, or both. As we become more concerned with land-use planning, it is important to recognize that the value of naturally subsidized ecosystems may result from their life-support and waste assimilation capacities as well as their potential for production. Any city that overtaxes its life-support module, or fails to preserve enough of it, soon finds itself caught in a vicious downward spiral of declining costs/benefits. The cost of paying for what was once the "free work of nature" soon overrides the benefits of life in the city.

One reason for our lack of awareness of the interdependence of fuel-powered and sun-powered ecosystems and the absolute dependence of the city on its countryside is economic. Up to now, the cost of energy, water, and so on has been low since these resources were supplied by nature with comparatively little cost to man. Now it is evident that we must place a more realistic economic value on the life-support environment so as to include the work of nature, if for no other reason than to remind us all of the true value of the self-sustaining solar-powered ecosystems, a value not now adequately reflected in its real estate price.

Production, Maintenance and Environmental Law

Eugene P. Odum

O ne of the major principles relating to the development of ecological systems has to do with the distribution of energy in the system. When the ecological system is young, the major flow of energy is directed to production, that is, to growth and the building of a complex structure. Organisms adapted to the youthful system are those with high birth rates, rapid growth potential, and the ability to exploit unused resources. However, as the population density approaches the saturation level, the ecological system matures in the sense that a greater proportion of the available energy is shifted to *maintenance* of the complex structure that has been created. Organisms adapted to the mature system, then, are those with low birth rates, longer life spans, the ability to recycle and reuse resources, and the capability of developing mutualistic relationships with each other. A parallel, of course, exists in the development of human society — one that is justified because man and environment do constitute an ecological system. Whether we like it or not, society is faced with the same problem nature has faced for millions of years; namely, making a transition from a transient, youthful state to a mature (and we hope enduring) state.

It is more than coincidental that we begin a new decade with mounting concern for human rights and environmental quality along with increasing unrest among the young and those who maintain our highly complex and

---

Originally published as the Foreword to H. Floyd Sherrod, ed., *Environmental Law Review*, Clark Boardman Company, New York, 1970.

technological society. As an ecologist, I view all of these trends as part of a perfectly natural and predictable expression of the basic need to develop new strategies adapted to the mature system. Up to now, the greatest economic rewards and the strongest legal protection have been given to those who produce, build, and exploit nature's riches; this, we can argue, is quite proper in the pioneer stages of civilization, since man must first to some extent subdue and modify his environment in order to survive in it. Now it is obvious that at least equal rewards and protection must be given to those people, professions, and industries that maintain the quality of human existence; survival in the future depends on finding a balance between man and nature in a world of limited resources. As this transition occurs, the basis for economic development shifts from exploitation to recycling, reusing, and prescription of the quality of resources (and man, too).

Legal procedures and legal education must adapt accordingly since the law, backed by strong public opinion, is the chief "negative feedback" that establishes the necessary controls. The traditional "private client-oriented" law must now be broadened to include greater emphasis on public and environmental law. University law schools, which have tended to be isolated from other academic schools and departments, need to move out of the legal ivory tower and establish better communication links with the environmental and social sciences, and encourage their students to seek better training in these and other relevant subjects. A first step in this educational transition, of course, is to make an inventory of environmental law as it is now understood and practiced. I am excited about work that brings together many of the highly specialized areas of law that are little known either to the public and the scientist, or, for that matter, to the average lawyer. I believe such work points up the urgent need to develop more comprehensive procedures that will counter the excessive fragmentation and help resolve the contradictions that now make it so difficult to deal with such problems as pollution on a legal basis.

Ecosystem Management: A New Venture for Humankind

Eugene P. Odum

*T*echnological assessment on a one-problem/one-solution basis will not be adequate to maintain a quality environment for humans. No matter how skillfully we monitor and deal with air pollution, water pollution, visual pollution, urban planning, economic development, and so on, as separate problems, the fact remains that it is the balanced interaction of the low-energy, sun-powered, self-nourishing (autotrophic) natural environment and the high-energy, fuel-powered, parasitic (heterotrophic) urban-industrial environment that must be the ultimate basis for environmental quality. Therefore, ecosystem management, which we define as assessment and management of humans and environment as one system, must rapidly replace current fragmented applied science that by and large attempts to manage components as separate entities.

Experience with insect control in cotton is a good example of the short-comings of the "quick-fix" approach, as was documented by Adkisson et al. (1982). Massive aerial spraying of insecticides in Texas resulted in increased yields for a number of years; but in the early 1960s an almost complete failure of the crop occurred, as the pest species had become resistant and other species of insects moved in to become new pests. Yields in this area

---

Previously published in Nicholas Palunin, ed., *Perspectives on Ecosystem Theory and Application*, John Wiley & Sons, New York, 1986.

were restored only after a more holistic approach, which entomologists term "integrated pest-control," was adopted. Experiences such as those are rapidly making it evident that complex situations and problems cannot be success-fully dealt with in the long term by focusing on only some component or components of the system. Accordingly, there is renewed interest in deter-mining in detail how ecosystem theory can help mankind to learn to coexist in proper harmony with the biospheral life-support system on which we ultimately depend.

Reasons for Slowness in Application

Before we review an updated theory of ecosystems, let us inquire into the reasons why "applied" scientists have been so slow to apply ecosystem theories to practical problems. The first and most obvious reason is that the "quick-fix" so often works very well in the short run of political and economic worlds, as in the cotton example. But when numbers of small "quick fixes" are made independently, the central problem is not properly addressed; decisions at a higher hierarchical level, that would benefit the whole, are accordingly not made. Economist Alfred Kahn has termed this situation the "tyranny of small decisions" (Kahn, 1966; cf. W.E. Odum, 1982).

A second reason, I believe, for delay in the widespread application of ecosystem theory is the fact that science has become so strongly reductionist that we are victimized by "a tyranny of small technologies." Although the philosophy of science has always been holistic, in the sense of attempting to understand the whole of the situation, the actual trend of science in the past few decades has become more and more reductionist — with increasing specialization, emphasis on smaller and smaller units down to the molecule and beyond, and a preoccupation with laboratory study. Much good, of course, has come from the reductionist approach. Mankind, in general, has benefitted from advances in medicine, engineering, and agriculture that have come from narrowly defined laboratory, greenhouse, and experimental sta-tion research. The real world, however, consists of open, far-from-equilib-rium, thermodynamic systems that cannot be enclosed in glass test tubes or within laboratory walls. Such open systems must have a strong input flow of high-quality energy to enable them to survive, let alone evolve and improve. They are also very much influenced by economic and political considerations that are rarely included in the scientists' models. For these reasons, real-world systems cannot be dealt with one piece at a time, or in isolation from economic and political co-acting systems.

A third reason why applied scientists have been slow to apply an ecosystem concept to practical problems lies, I believe, in the mistaken notion that the whole is nothing more than the sum of its parts. According to this widely held philosophy, the whole can be understood on the basis of detailed study of major components. In a book on experimental biology, Norman (1963) expresses this viewpoint as follows: "The whole living organism is nothing more than the sum of its parts. There is nothing more about the life of the whole organism that cannot be explained by the physics and chemistry of the individual activities." Then Norman attempts to justify the experiments on parts of organisms rather than the whole organisms by claiming: "Parts respond the same when isolated as when contained in the intact organism [so that] once the activities are isolated, and parts are understood, it will be possible to integrate these understandings into an explanation of the life of the intact cell."

By inference, these beliefs regarding the cell and the organism are transferred in the minds of the great majority of physical scientists and biologists who focus on the physics and chemistry of life as meaning that the larger systems, such as populations and communities, can be understood by the study of isolated species or other components. Ernst Mayr, in his book *The Growth of Biological Thought* (1982), argues strongly for the proposition that biological systems differ from physical systems in the relationship between parts and wholes.

One historian, Lynn White (1980), believes that an excessive reductionist tendency in science leads to a decline in public support of science: as scientists become more and more specialized and introverted, they tend to be less and less able to deal with the macro scale of real-world problems.

Emergent Properties

The ecologists, and also philosophers, have for a long time maintained that, at levels of organization above the organism, the whole is more than a sum of the parts. In other words, as components are combined to produce larger and more complex systems, new properties, often called "emergent properties," appear.

In my new book (E.P. Odum, 1983) I cite two examples, one from the physical realm and one from the ecological realm, to illustrate emergent properties. When hydrogen and oxygen are combined in a certain molecular configuration, water is formed — a liquid with properties that are utterly different from those of its gaseous components. Likewise, when certain plants (algae) and coelenterate animals evolve together to produce a coral, an efficient, nutrient-cycling mechanism is created that enables the combined system to maintain a high rate of productivity in waters with very low nutrient content (see Muscatine and Porter, 1977). Similarly, mutualism between mycorrhizae and trees enables tropical rain forests to prosper on infertile soil (see Jordan, 1982). Thus, the fabulous productivity and diversity of coral reefs and rain forests involve emergent properties that are only found at the level of the community.

We can say in all truth that scientists of many different disciplines are badly split on the matter of reductionism versus holism. The difficulty of dealing simultaneously with the part and the whole is also reflected in the conflict between the individual "good" and the public "good," which has for many centuries confounded philosophers and bedeviled society. Numerous economic and political approaches, designed to deal with this conflict, have been suggested or tried — but as yet with little success. In the United States, as well as in many other countries, elected governments over the years have shifted back and forth from strong attention to the individual (the conservative stance) to an emphasis on public well-being (the liberal stance). As a result, the parts (individual) and the whole (public) get attention, but not at the same time. The short-term nature of most political and economic procedures makes it very difficult to integrate these two concerns. I believe that the study of how natural ecosystems operate and evolve may help us to resolve this basic human problem.

The relationship between the parts and the whole may well depend on the level of complexity of the system under consideration. At one extreme, ecosystems that are subject to severe physical limitations (as on an arctic tundra or in a hot spring) have relatively few basic components; in such "low-numbers systems" the whole is indeed probably very close to the sum of the parts, having few, if any, emergent properties. In contrast, "large-numbers systems," such as The Biosphere, have a great many components that act synergistically to produce many emergent properties; in them the whole is definitely not just the sum of the parts. Indeed, studying all of the parts separately in such a complex system is completely out of the question, so one must focus on the properties of the whole.

Most ecosystems, and many of our day-to-day problems, as delimited in practice, are "middle-numbers systems." Allen and Starr (1982) have discussed such "middle-numbers systems" and have concluded that the best approach to their study and management is a hierarchical one. The assumption is that complex systems are hierarchical in function and structure, and therefore are best understood in terms of levels of organization within the total system. In this approach, one first delimits an area, a system, or a problem of interest, as a sort of "black box." Then the energy and other inputs and outputs are examined, and the major functional processes of the system as a whole are assessed. Following the parsimonious principle, one then examines such operationally significant components or groups of components (populations and physical factors) as are determined by observing, by modeling, or by perturbing, the system (as a means of identifying operationally important components). In this approach, one goes into great detail in the study of components, but only as far as may be necessary to understand or manage the system as a whole.

To Bridge the Gap Between Theory and Practice

Two aspects of ecosystem theory, as indicated in Figure 8.1, need to be emphasized if we are to bridge the gap between theory and practice. The internal dynamics of an ecosystem — involving the flow of energy, the recycling of materials, the organization of food-webs, and so on — have been much studied and are fairly well understood. In contrast, what might be called the external dynamics of an ecosystem have been widely neglected. As ecosystems are open systems, consideration of both the input environment and the output environment is an important part of the concept. These environments are coupled with, and essential for, the ecosystem to function and maintain itself. The left-hand diagram in Figure 8.1 emphasizes this feature. The conceptually complete ecosystem, accordingly, must include an input and an output environment, along with the system that is being delimited studied, or managed. Accordingly, Ecosystem = IE + S + OE. For more on the concept of input-output environment, see Patten (1978) and Patten and Odum (1981).

Viewing the ecosystem in this manner solves the problem of where to draw the lines around an entity that one wishes to consider. It does not matter very much how the "box" portion of the ecosystem is delimited, provided the inputs and outputs are included. Often natural boundaries, such as lake shores, forest edges, or political lines (such as city limits), make convenient boundaries; but limits can just as well be arbitrary, so long as they can be accurately designated in a geometric sense. Too often, ecosystems are conceptualized and modeled as closed systems, as if the box part in the diagram in Figure 8.1 was an impervious container. If this were so, the living contents

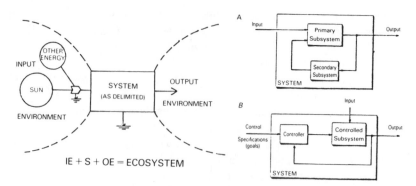

Figure 8.1 Two important aspects of ecosystem theory. A real ecosystem, however delimited, is an open, far-from-equilibrium, thermodynamic entity requiring strong inflows and outflows. Accordingly, an ecosystem is the sum of the input environment (IE), the area with its biota as delimited (S), and the output environment, as shown in the left-hand diagram (after E.P. Odum, 1983). The right-hand diagram contrasts the cybernetics of natural ecosystems, A, with that of man-made or organismic systems, B. The latter have specific controllers and set-points or goals, while natural ecosystems are controlled by feedback from numerous, diffuse subsystem networks (after Patten and Odum, 1981).

would not survive such enclosure. A functional real-world ecosystem must have an input lifeline and an output of processed energy and materials.

The extent of the input and output environment (and, therefore, the importance of including them) depends upon a number of variables, for example, (1) the size of the system: the larger the system is, the less will be its dependence on externals; (2) metabolic intensity: the higher the rate is, the greater will be the input and output required to sustain it; (3) the balance between autotrophs and heterotrophs: the greater the imbalance is, the more will be the externals required to maintain balance; (4) the stage of development: young systems differ from mature systems in that the former are more influenced by the input environment than are the latter.

To illustrate some of these differences, we might compare a large forested tract with a city of the same area. Both are very open systems, but the city has a very much larger input and output environment than the forest, because the former is essentially a very high-powered heterotrophic system that depends on huge inflows of energy and materials, with correspondingly large outputs of processed energy, materials, and wastes. What we are saying here is that an ecosystem can neither be understood by taking it apart and studying the parts piecemeal, nor can it be dealt with by conceptually enclosing it.

Considering the input environment system can be illustrated by a comparison of two bays on the Florida panhandle, as described by Livingston (1980). One bay is the mouth of a large river and is thus "river-driven." Allochthonous detritus, imported by the river, dominates the community metabolism. The water is turbid and highly colored. Mud-flats and productive oyster reefs abound, and seasonal pulses are related to rainfall and river flow. In contrast, the input to a neighboring bay consists largely of gentle tidal inflows from the open sea. The water is clear and the bottom dominated by macrophytes (seagrass beds). Phytoplankton production is low and runs on internally regenerated nutrient. The input environment is thus smaller and has less influence than in the case of the river-driven bay. Both bays are subjected to periodic storms that are followed by some reorganization of successional sequences. Both the continuous and periodic inputs must be primary considerations in any management decisions regarding oyster, shrimp, or any other component.

Control Mechanisms in Ecosystems

Another aspect of ecosystems that needs clarification involves the type of control mechanisms. It is very important to recognize that the cybernetics of ecosystems is rather fundamentally different from the cybernetics of organisms and man-made engineering systems, as is illustrated by the right-hand diagrams in Figure 8.1. In both engineering (servo-mechanism) and organisms, a distinct mechanical or anatomical "controller" has a specified "setpoint" (diagram B). In a familiar household heating system, the thermostat controls the furnace. In a warm-blooded animal, a specific brain center controls the temperature. In contrast, it is the interplay of material cycles and energy flow along with subsystem feedbacks that control large ecosystems. Ecosystems thus have self-correcting homeostasis via feedback from subsystems, but no outside controls or goals (diagram A). Ecosystems are not goal-oriented but are self-organizing and controlled. In addition to feedback control, redundancy — that is, more than one species or component capable of performing a given function — also enhances stability in ecosystems.

One difficulty in perceiving cybernetic behavior at the ecosystem level is that components of the ecosystem are coupled in networks by various physical and chemical messengers that are analogous to, but far less visible than, nervous or hormonal systems of organisms. Simon (1973) has pointed out that "bond energies," which link components, become more diffused and weaker with an increase in space and time scales. At the ecosystem scale, these very numerous bonds of energy and chemical information are the controlling factors. Organisms responding dramatically to low concentrations of substances, or low populations of strongly interacting organisms, are more than just a weak analogy to hormonal control. Low-energy flows producing high-energy effects are ubiquitous in ecosystem networks. The quality of energy, and the topographic position within networks, are often more important than the quantity of energy. For example, parasites and predators may account for only a very small portion of the total community metabolism, yet by their impact on herbivores they can have very large and even controlling effects on total primary energy-flow. Since 1970, there have been a number of excellent experimental studies demonstrating that so-called "downstream" components can have both negative and positive feedback effects on "upstream" components. This type of amplified control, by virtue of position in a network, is exceedingly widespread and indicates the intricate global feedback structure of ecosystems. Through evolutionary times, such interactions have stabilized the ecosystems by preventing boom-and-bust herbivory, catastrophic predator prey oscillations, and so on. The Gaia Hypothesis, as presented by Lovelock and Margulis (see Lovelock, 1979), is based on the concept that the entire biosphere is controlled by very numerous but very diffuse subsystem interactions.

The challenge in working at the ecosystem level is to identify and measure not only the large and conspicuous biomass structures, flows of energy, and cycling of materials, but also the key high-quality but low-energy flows that constitute controlling subsystems. In forests, for example, conventional methods of assessing net primary production that involve periodic harvest of living and dead plant tissue completely overlook the outflow of dissolved organic matter from leaf and root exudate and the photosynthate extracted by mycorrhizae. The latter greatly augment the inflow of nutrients to the plant. A mycorrhizal network not only enhances overall primary productivity, but also serves as a diffuse controlling subsystem (Odum and Biever, 1984). Accordingly, toxic wastes, soil disturbance, or other perturbations could have their greatest effect on the ecosystem as a whole through disturbance of microbial networks — components that are rarely considered when impact studies are carried out.

Ecosystem Concepts and Management

Although forests, as ecosystems, are not completely understood, ecosystem concepts are being successfully applied in forest management (e.g., Waring, 1980). Regional models that integrate allogenic (forcing functions in the input environment) and autogenic (occurring within the bounded system) processes are proving useful in predicting future forest composition and timber production. For example, in the Piedmont region of Georgia, natural autogenic secondary succession proceeds from old fields to pines and thence to hardwoods. Commercial forestry as an allogenic force, however, works to increase and maintain pines and discourage conversion to hardwoods. Johnson and Sharpe (1976) used inventory data for the previous ten years to model trends for the next twenty-five years. Results indicated that, although future compositional dynamics will be strongly influenced by man's efforts to maintain pines, natural succession combined with increasing urbanization and fire suppression (all three of which favor hardwoods over pines) will predominate, so that the area in pines will decrease and the area in hardwoods will increase.

The contrast between a piecemeal and a holistic view of land use is illustrated by the piecharts in Figure 8.2. In each diagram the landscape is subdivided into three basic systems: urban, agricultural, and natural (which represents in a broad sense the life-support module of our Spaceship Earth). In the upper diagram the percentage of area actually occupied by each is shown. As the power density (i.e., energy-flow per unit area) and energy quality (fossil fuels versus solar) required differ so greatly in the three systems, this diagram is very misleading. Although urban-industrial systems (including transportation corridors in rural areas) occupy only 6% of the land area, their relative importance is very much greater, due to those huge demands of energy and materials that have to be drawn from the other two systems — not to mention equally large outputs of wastes that impact on those other areas. The upper diagram gives the very false impression that there is plenty of room to expand the urban and agricultural area more or less indefinitely. Current political and economic decisions for the most part are carried out as if each of the three systems was an independent rather than an interdependent entity.

A more realistic view of our total landscape is obtained if we weight each subsystem according to energy density, as is shown in the lower diagram of Figure 8.2. If we consider that the urban ecosystem is at least ten times, and the agroecosystems two times, as "energetic" as the natural ecosystems, then on a percentage basis each component has an approximately equal impor-

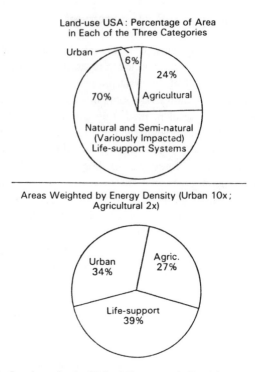

Figure 8.2 Land use in the United States symbolized in terms of actual areas occupied by major system types (upper diagram) and in terms of areas weighted for energy density (lower diagram).

tance value. The 10× and 2× weighting factors are very conservative and do not take into consideration energy quality differences. Whatever the factors, it is certain that for every hectare of highly developed landscape many hectares of natural and semi-natural environment are required for life support.

It can be said that the environmental quality of the United States (as indicated by air and water quality, recreational opportunities, etc.) is generally good but beginning to show distinct signs of deterioration. Accordingly, we might conclude that current partitioning of the landscape is a favorable one, and further expansion of urban development must be accompanied by increasing efficiency of energy, water, and material uses and a decrease in waste outputs. Otherwise, all three interrelated systems will degrade simultaneously. There is much discussion nowadays about "endangered species," but what is really endangered and surely matters most is our and their life-support system!

The carrying capacity of the biosphere for humans is unknown, and may be nearly impossible to estimate on any kind of scientific basis. We shall presumably have to continue to depend on indicators such as air pollution, food shortages, etc., to warn us when our life-support module is unduly stressed. To take any kind of effective action to reverse an undesirable trend would require that we move as rapidly as possible to ecosystem-level thinking and management. This needs, among other prerequisites, the development of a "holoeconomics" that integrates market and non-market values. I personally believe that a favorable future for mankind depends more on the integration of ecology and economics than it does on new technology, although we will need that as well.

Summary

Recognition that complex systems and situations cannot be understood or dealt with piecemeal leads logically to consideration of the ecosystem level of organization. At this level, major properties and processes result, not from summation, but from integration and co-evolution of biotic communities and abiotic environments. Ecosystems resemble organisms in being open, far-from-equilibrium thermodynamic systems with input and output environments; but they differ from organisms in the way in which they develop and are controlled. A holistic or holoeconomic approach is especially important in assessing land use. Only in this manner can non-market life-support goods and services of natural environments be properly valued and effectively preserved.

The Watershed as an Ecological Unit

Eugene P. Odum

The first statement that we might make is that man's success in manipulating his environment is proportional to the degree to which the whole environment is considered in formulating the strategy. Or, to put it another way, mistakes too often result from overemphasis on the kind of management that produces a short-term benefit to the part, but a long-term detriment to the whole of the environmental system. The tendency to overuse fertilizers and insecticides in agriculture is a good example. So long as an increase in crop yield is the only consideration, and the effect of the cropping procedure on the watershed system is ignored, then it is inevitable that there will be trouble downstream. As you know, Lake Erie is in trouble, but not because of anything anybody did to the lake itself. The lake is severely stressed because of all the stuff that's going into it from the watershed. It is the excess fertilizers from the agricultural drainage and the untreated wastes from the cities that are causing the "cultural" eutrophication. We hear a great deal about America's efficient agriculture and industry, but if we "tell it like it is," as our young people would have us do, then we must own up to the fact that our agriculture and our cities are grossly inefficient in terms of the basic

---

Reprinted from *The Cuyahoga River Watershed*, proceedings of a symposium commemorating the dedication of Cunningham Hall, Kent State University, 1 November 1968. Sponsored jointly by the Department of Biological Sciences, Institute of Limnology, and the College of Arts and Sciences, Kent State University, and the Three Rivers Watershed District and the Lake Erie Watershed Conservation Foundation, Cleveland, Ohio.

ecological necessity for recycling of materials. Otherwise, what are all those fertilizers and pesticides doing in Lake Erie? If we were really efficient in applying chemical additives to croplands, the fertilizer would mostly end up in the crops and the pesticides would not leave the site where they are supposed to control the pests. The mass chemical approach to the control of pests is turning out to be a sad example of short-term gain followed by long-term disaster. The development of resistance by pest species and the destruction of natural enemies combines to wipe out the advantage in just a few years. We then return to the "old-fashioned" strategy of a mixed bag of cultural, biological, and chemical controls, which, to save face for conceited mankind, is given the new name of "integrated control"!

Agricultural drainage, of course, is only one contributor to imbalances in water. A rather different crisis is caused by the industrial outpourings of substances that are new or foreign to ecosystems and, consequently, often toxic or nondegradable. Again, the short- and long-term motivations stand in startling contrast. Long-chained or "hard" detergents and nonreturnable aluminum beer cans appeal to the user on a short-term basis, but they prove to be very expensive to him in the long run, because he must ultimately suffer increased taxes to pay for the extra or special waste disposal systems for the undecomposable items. The old glass bottles worth 5¢ on return were never an environmental problem because there was an economic incentive to recycle! And many a kid profited from collecting bottles along the roadside. A great deal of pollution and costly waste disposal could be avoided if recycle and waste disposal problems were anticipated in product design, but ignorance, greed, and public apathy make it difficult for the industrialist to consider the whole cycle of production and utilization. Murphy, in a book entitled *Governing Nature* (1967), emphasizes that economic incentives must accompany regulatory restriction if pollution is to be controlled. I would add that economic incentives or pressures should apply both to the consumer and the manufacturer. Let us be fair about this and place the blame on both parties that conspire to ruin our environment!

In times of rapid change or stress, we often resort to slogans, or battle cries, as a means of obtaining a directed response from society. "Too much of a good thing" might serve as an appropriate slogan to dramatize a way out of the environmental crisis. Up to now, man has been concerned with not having enough. Throughout our history we have had to be preoccupied with striving for more in order to establish ourselves as a successful species on this earth. First, it was more land in cultivation, then more people, and now it is more automobiles and more urban sprawl. In the early development of civilization, rapid growth and exploitation of resources are desirable attributes or "good things," so to speak. But, sooner or later, the saturation level is approached and we must be concerned about too much of these "good things." The problem is that society has not yet developed an effective political–legal mechanism that is able to detect and counteract excesses. Environmental sensors and a central control "thermostat" either do not exist or are inadequately linked by circuitry, largely because it is understandably difficult for all of us to accept the fact that the very things that have made us rich and prosperous can now literally choke us to death if we do not establish rational plans for control. The watershed concept provides an immediate focus for overcoming this "cultural lag," because the water cycle is a key circuit in the environment that is easy to measure. It is one that can be readily understood by the public. Furthermore, the drainage basin is an ecological system large enough for the practical exercise of integrative study and control, yet not so large as to include too many uncooperative political units. Another beauty of the watershed idea is that it is easy to visualize; one can draw a discrete line around it on a map. As professionals we can argue about the hydrological basis to be used in the exact placement of the line, but at least Mr. Citizen can understand the unity of a drainage system.

The watershed as a unit for study and experimentation is, of course, not a new idea. Until quite recently, watershed studies have been quite narrow in that attention was sharply restricted to the behavior of water itself, or to the control of soil erosion. At the Coweeta Hydrological Laboratory in western North Carolina, for example, the input of rainfall and output of stream flow has been continuously monitored for many years on watersheds subjected to different types of forest and other land management. Hibbert (1967) has summarized some of the long-term results of these studies. Moderate reduction in the volume of vegetation can result in increased water yields from forested slopes without damage to soil, provided timber removal procedures are designed with care. Pine forests lose more water by evapotranspiration and hence dry out the land and reduce stream flow to a greater extent than deciduous forests, apparently because of year-round interception of rainfall and subsequent evaporation from foliage. Studies on another experimental watershed, the Hubbard Brook Experimental Forest in New Hampshire, are showing that gains in water quantity come at the expense of water quality; deforestation results in increased nutrient content of the runoff water, which may then contribute to downstream eutrophication (Likens et al., 1969).

The watershed as a unit for planning is also not new. Again, however, the mission orientation has been far too limited and fragmented. Most so-called watershed "development plans" turn out to be engineering exercises in the construction of impoundments that are sold to the uninformed public as cure-alls for all ills and as the bonanza for all uses. In actual practice, carefully planned impoundments are useful additions to the total watershed system, but are ineffective in curing many ills such as pollution that origi- nates in the watershed itself. Money spent on large dams might better be diverted to rebuilding cities and to abating pollution at its source.

If the watershed is to be a really effective unit from the standpoint of the total ecology of man and environment, then, in my opinion, three drastic changes in terms of approach and policy are necessary.

First, watershed studies must be extended to include input–output budgets and rate of cycling measurements of nutrients as well as of water. More emphasis needs to be placed on the biotic components that ultimately control vital nutrient cycling. One of the principal objectives of the American plan for the International Biological Program is to promote such an ecosystem approach to the study of watersheds in each of the major biomes of North and South America.

Second, the strategy of "multiple use" must be replaced by a strategy of "compartmentalization." Elsewhere (Odum, 1969a,b), I have discussed in detail the basic ecological theory that must be considered if we are to resolve conflicts, and the consequent need for zoning the non-urban as well as the urban landscape. Figure 9.1 is a compartment diagram illustrating the four kinds of environments required by man. A simple illustration will perhaps clarify the principle. Fishing and water supply for cities are conflicting uses for a lake or impoundment. An infertile or oligotrophic lake not subject to algal blooms is ideal for water supply for a city, while a fertile or eutrophic lake is needed for fish production. One cannot maximize for both in the same lake. Since a compromise brings on conflicts, it would be desirable to compartmentalize rather than compromise wherever possible. Two lakes, each "zoned" and "managed" for the different uses, would be infinitely more satisfactory and would contribute to the stability of the overall environment.

Finally, planners must become more aggressive in the use of legal and political authority to make certain that watershed plans are actually carried out and are not overturned by short-term pressures. Education to establish strong public opinion is, of course, vital to the exercise of power. Environ- mental rights must become a part of human rights.

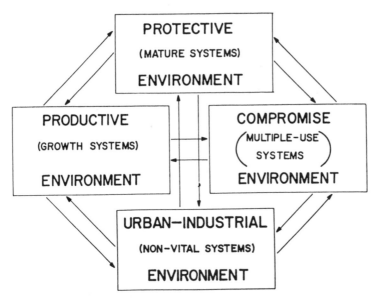

Figure 9.1   A compartment plan for a large area of the countryside such as a watershed unit. Each of the four types of environments shown represent different types of ecological systems that must function together if quality rather than mere quantity of human existence is to be the long-range goal of society. Since each ecosystem type requires a different management strategy, some kind of zoning or use restriction procedure is necessary. The model emphasizes that a mature forest that protects an urban water supply is just as important as productive croplands, and that not all the landscape can be "comprised" by multiple use. It must also be recognized that cities cannot be allowed to expand indefinitely at the expense of the ecological systems that support them.

In summary, it is clear that the watershed is a unit that can serve as a large "eco-stat" for maintaining a safe balance between needs for "protection" and "production." Whether the watershed is a suitable unit from the socioeconomic standpoint is to be discussed in subsequent papers in this symposium. It may boil down to a practical judgment of size. Many natural watersheds may be too small. Combining drainage basins on the basis of urban patterns, as is the case in the "Three Rivers Watershed District," seems to me to be a splendid idea.

Natural Areas as Necessary Components of Man's Total Environment

Eugene P. Odum*
and
Howard T. Odum**

O ur theme is that the natural environment is an essential part of man's total environment. Preservation of a substantial portion of the biosphere in a natural state, while not a panacea for all the ills of mankind, is nevertheless a necessity if we base the carrying capacity of the earth on the quality of human life. First, we define "natural environment" as that part of our environment that is essentially self-supporting, in that a minimum of human management is required for maintenance. In terms of function, "natural environment" is that part of man's life-support system that operates without energetic or economic input from the power flows directly controlled by man. "Natural environment" is a more restricted category than

Reprinted from *Transactions of the 37th North American Wildlife and Natural Resources Conference*, March 12–15, 1972. Published by the Wildlife Management Institute, Wire Building, Washington, DC 20005.

*University of Georgia, Athens.

**University of Florida, Gainesville.

"open space," a term widely used by planners to mean any part of the landscape, whether natural or man-made, that is, free of building structures. In this context, "natural environment" includes ecological systems ranging from little-used wildernesses to moderately used forests, grasslands, rivers, estuaries, and oceans, which produce useful products and recycle wastes on a continuous basis, but *without appreciable economic cost to man*. These self-maintaining ecological systems run on sun energy, including the energy of rain, wind, or water flow that are derived from sun power. In contrast, what we choose to call "developed environment" includes ecosystems that are structured and maintained by large auxiliary power flows from fossil or other concentrated fuels that supplement or replace the natural energy flow of the sun. A city, of course, is the ultimate developed ecosystem, but golf courses, suburban developments, agricultural fields, and channelized rivers are also developed ecosystems since they require a diversion of energy from man-controlled power flows to maintain them in the developed state even though natural elements (water, trees, grass, bacteria) may have important roles in such systems. Developed systems generate economic wealth, but *the economic cost of maintenance increases as a power function of the intensity of development*. For example, it is well known that the cost of maintenance (C) of a network of services increases roughly as the square of the number of units (N) in the network, as shown in the following equation:

$$C = \frac{N(N-1)}{2}, \quad \text{or approximately} \quad C = \frac{N^2}{2}.$$

Thus, if a city doubles from 10 to 20 million units, the cost goes up 4 times. Furthermore, the stress on supporting natural life-support systems increases markedly, again as some kind of multiplier, as the size and power demand of developed systems increases. *Because the multiplying maintenance costs are too often not anticipated and the useful work of nature totally undervaluated, developed systems have an inherent tendency to grow beyond optimum size, and at the expense of natural systems.*

In some parts of the world, aesthetic and recreational values (and associated economic dividends) have been sufficient to justify preservation of large natural areas in parks and refuges. In some countries, preservation of greenbelts and other natural areas have been a cultural or religious tradition. However, in the future neither aesthetic values nor ethnic traditions will be adequate bases for preservation of natural environment, because rapid technological and population growth produces a strong drive to convert natural environment into developed environment. General public awareness that natural environment is important is, by itself, not enough. So powerful is the positive feedback within the urban system, and the economic "forcing function" from outside, that there has to be equally strong negative feedback control built into economic and political systems to prevent overdevelopment. To suggest that cities and other highly developed ecosystems have an inherent tendency to grow beyond the optimum (i.e., to "overdevelop") is not to embrace an anti-human, anti-urban, or anti-development philosophy. Because cities and other developed environments are so valuable to man, they must be protected from exploitation, just as is necessary for any valuable resource. Specifically, cities need the protection of an adequate life-support system, many elements of which natural environment provides free of charge. Without natural recycling and other work of nature, the cost of maintaining quality life in cities would be prohibitive. Later in this paper we will show by actual calculation that the per-capita cost of treating human wastes, which are only one small part of the pollution disorder generated by cities, would be more than doubled if there were no natural environment available and able to carry out the work of tertiary treatment of these wastes.

A first step towards redressing the imbalance in valuation of natural versus developed environment would be to determine the real value of "natural environment" in comparable monetary terms as are used to determine the worth of developed environments. The example of the previous paragraph suggests one approach, and we will have more to say about this later in this paper. However — and this is our most important theme — *the true value of a man's total environment is determined by the diversity interaction between the "developed" and the "natural" environment and not only by the worth of each as a separate component.* Yet, at the present time society does not evaluate in any effective manner total environment, but bases human values on the monetary worth of separate components, largely the highly developed ones. If power-hungry developed systems spread in an unrestricted and unplanned manner at the expense of the natural environment, then a point is soon reached where the latter is unable to perform its "free" life-support functions. Then, not only does the quality of the remaining natural self-supporting environment decline, but, more important, the quality of the highly developed environment also deteriorates as the costs of pollution and other disorder abatement rises precipitously in nonlinear, multiplying fashion. *Accordingly, there has to be some optimal proportion between the natural and developed environments* (since 100% of either would be unthinkable). Once a rational ratio for a given region is determined, there has to be an agreed upon "environmental-use plan" (equal to a "land-use plan," as this term is generally understood by planners and conservationists), with sufficient legal and political sanction to counteract the overdevelopment syndrome. We aim to show that it should now be feasible to model environmental decision-making so as to predict the total consequences of varying the proportion of developed to natural environment (1:1, 2:1, and so on) and thereby find an optimum range in terms of quality of the total environment. After we have presented a very simplified and theoretical working model for such ecosystem management, we will then discuss more pragmatic approaches that we hope planners will find useful until such time as realism can be built into total models.

The Ecosystem Management Model

Figure 10.1 pictures the essential elements, energy flows, and human values that must be considered in modeling environmental-use options designed to maintain an optimum balance between natural and developed ecosystems in counties, watersheds, states, regions, or other relatively large areas of the biosphere. The symbols used in this modern part of an "energy language" were devised by H.T. Odum and are described in detail elsewhere (Odum, 1967, 1968, 1971). Circles represent energy sources, while the tank-shaped round bottom modules represent stored or potential energy or resources. Natural ecosystems are depicted as bullet-shaped autotrophic modules that are self-nourishing, and developed ecosystems as hexagonal, heterotrophic, or consumer modules that require an energy input for nourishment. Both natural and developed ecosystems have storage capacities important for maintaining function during periods of reduced energy inflow. Downward-directed arrows into heat sinks (like electrical "ground" symbols but only one-way flow) show where energy is lost during conversion from one form to another, as required by the second law of thermodynamics. Modules containing a large X stand for a multiplicative function during work transfer or exchange, as discussed in the preceding section of this paper. Especially important is the energy drain and reduction in storage capacity that developed systems impose on natural systems, as shown by the "stress" arrow in Figure 10.1. Note that, except for original energy sources and land resources, all modules have at least one input, or source of energy, and at least two outputs, one representing a heat loss or energy drain, while the other is passed on as an input to another unit in the system. Seen graphically in this manner, the interrelationships and interdependence of components as working parts of the whole can be clarified. Finally, the model identifies the three environmental values as previously mentioned, namely: (1) the value of natural ecosystems as such, (2) the value of developed ecosystems as such, and (3)

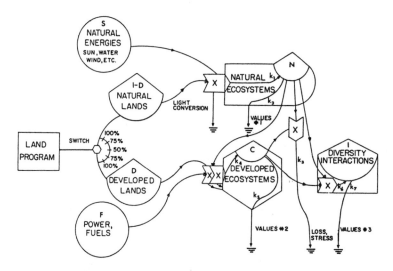

Figure 10.1   A model for land management in which the proportion of natural and developed lands can be varied in order to determine the optimum balance in terms of value of the total environment (sum of values 1, 2, and 3).

the value of man's total environment as determined by "diversity interactions" between 1 and 2.

If all of the inputs and outputs of components in the ecosystem model can be quantitated in common-denominator energy units (which can be converted to monetary units, as will be discussed in the next section), different options in land-use planning can be simulated with analog or digital computers. The switching module in Figure 10.1 shows how a land-use program could theoretically vary the percentage of land that is developed in order to predict interactions, and to determine an optimum proportion between developed and undeveloped land in terms of the quality of the total environment (value 3, Figure 10.1). To be realistic, the oversimplified model of Figure 10.1 would have to be expanded to include separate modules for different types of both developed and natural systems, since energy requirements and outputs vary widely within each of these two general classes of ecosystems. For example, as already noted, a high-density city has a much greater power requirement, and therefore exerts a much greater stress on its surrounding environment, than does a low-density suburban development.

Since the metabolism of a modern city, with its automobiles, industries, and high electric power consumption, is about 100 times greater than the metabolism of most natural ecosystems (4000 as compared to 40 kcal/m²/day), it is easy to see why high-powered systems tend to be destructive of the lower-powered systems in contact with them. Even the simplest models clearly demonstrate that high-powered systems, such as cities, require abundant life support from nature. If large areas of natural environment are not preserved to provide the needed input from nature, then the quality of life in the city declines and the city can no longer compete economically with other cities that have an abundant life-support input. Frequently, it is not energy itself that becomes the limiting factor, but some basic natural resource required to maintain the high rate of energy flow. In South Florida, water seems now to be that limiting input. Continued urban or industrial growth in many areas will depend on developing special water sources such as by desalination or by pumping water from underground or distant sources. If such special sources are developed, the city's energy cost will rise until it can no longer compete with cities that do not have to pay this extra cost. It is a sad situation when cities grow beyond their means and can no longer pay for their own maintenance. They borrow money or demand federal grants in order to grow ever larger and more demanding of their life-support system, when they ought to be diverting more of their energy to maintaining the quality and efficiency of the environment already developed, and to reducing the stress on vital life-supporting natural environment. Preliminary simulation of the South Florida situation indicates that a 1:1 ratio of natural to developed environment would provide a basis for an optimum environmental-use program. Until this kind of systems analysis procedure can be refined and become a basis for political action, it would be prudent for planners everywhere to strive to preserve 50% of the total environment as natural environment.

Calculating the Monetary Value of Natural Environment

As indicated in the preceding section, it will be a long time before total ecosystem management will be accepted as an economic and political reality. In the meantime, we have to justify and manage on the basis of separate values (values 1 and 2, Figure 10.1). A stronger economic basis for justifying the preservation of natural environment is obtained if we calculate the work of nature in terms of dollars or other currency units. Since money and energy flow in opposite directions, which is to say that money outputs is exchanged for energy input, H.T. Odum (1971) has suggested that the ratio of Gross National Product (GNP) to National Power Consumption can be used to convert calories to dollars. For the United States, this works out to be approximately 10,000 kilocalories equal to one dollar. Using this conversion, Lugo et al. (1971) calculated the work done by a tree with a 50 m^2 crown as being worth $128 per year, and $12,800 over the 100-year life span of the tree. The useful work done by an acre of forest, then, would be $10,360 per year and $1.04 million over a 100-year period. This value may be regarded as somewhat inflated by egocentric man, since he might not consider all work done by a forest useful to man. However, we believe it comes closer to the real value than conventional economic cost-accounting, which values a forest only in terms of yield of wood or other consumer products and ignores its life-support value.

Another approach to economic justification for preservation of natural environment involves evaluating the work of nature in treatment and recycling of wastes. Again, conventional accounting rarely includes placing a dollar value on such useful work. This can be done by calculating how much it would cost cities to completely treat wastes by artificial means if there were no natural environment available to do at least part of the work. Experiments at Pennsylvania State University have shown that land areas covered by natural or semi-natural vegetation can be effective natural tertiary treatment areas for municipal waste that have gone through secondary treatment (see Parizek, 1967; Sopper, 1968). While these studies suggest that 2 inches a week of wastewater can be added without stress, we would suggest that about half of this, or 4 feet per year, would be a more judicious rate in terms of avoiding mineral buildup in the land filter. An acre of land could then absorb about 1.3 million gallons of treated wastewater per year, which is about the amount of wastewater produced by 35 city people (100 gallons per day per person, or 36,500 gallons per year). If this waste were subjected to artificial tertiary treatment, the cost would be 30¢ per 1000 gallons or about $400 for the 1.3 million gallons. Thus, an acre of natural environment could be worth at least $400 per year for this one useful function alone. Most of all, if all wastes had to be carried through tertiary treatment in artificial systems because there was not enough natural environment to do this work for free, then the taxpayer's bill for waste treatment would be doubled since tertiary treatment costs about twice as much as secondary treatment.

The Per-Capita Approach

E.P. Odum (1970) attempted to determine the total environmental requirements for an individual as a basis for estimating the optimum population density for man. In this study, the State of Georgia was used as an input–output model for estimating the per-capita acreage requirements on the assumption that this state is large enough and typical enough to be a sort of "microcosm" for the nation and the world. The basic question asked was: How many acres of environment does each person require to maintain a reasonably high standard of living on a continuing self-contained equilibrium basis — in the sense that imports and exports of food, other energy, and resources would be balanced? In other words, what does it take to support a quality human being in an area that cannot count on being an ecological and economic "parasite" on some distant region. As it turned out, Georgia is a good microcosm for the United States because its human density, growth rate, food production, and the distribution of its human and domestic animal populations are all close to the mean situation for the whole nation.

The per-capita area required for food was estimated by taking the diet recommended by the President's Council on Physical Fitness and determining how much crop, orchard, and grazing land is required to supply the annual requirement for each item. If we could be satisfied with a diet based on intensive grain and soy bean culture, perhaps as little as one-third of an acre could keep a person fed and reasonably well nourished, but the kind of diet Americans enjoy (including orange juice, bacon and eggs for breakfast and steaks for dinner) requires a great deal of land to produce, at least 1.5 acres per person. The impact of domestic animals on man's total environment is often overlooked in land-use planning. In Georgia, for example, domestic animals (cattle, pigs, chickens, etc.) consume primary production (food produced by plants) equivalent to that consumed by 21 million persons (compared to 4.8 million persons now living in the state). And this does not include pets, which for the nation as a whole consume enough food to feed five million people. While the impact of domestic animals on the environment is not nearly so great as that of an equivalent human biomass, the stress they place on the natural environment is considerable, and must be accounted for. We could, of course, do away with domestic animals, but this would mean giving up meat in the diet (and associated options) and dehumanizing man himself to the level of a domestic animal.

In a similar manner, the per-capita acreage needed for fibers (paper, lumber, cotton, etc.), watersheds, tertiary treatment of wastes, recreation, parks, highways, and urban and industrial living space were estimated. For some uses good data are available in statistical yearbooks, but for other needs (for example, outdoor recreation) we had to depend on recommendations of professional planners who deal with the particular human need. Our preliminary attempt to sum up total environmental needs in terms of the minimum space required is shown in Table 10.1.

Food-producing land	1.5 acres
Fiber-producing land	1.0 acres
Natural use areas	2.0 acres
Urban-industrial systems	0.5 acres
Total	5.0 acres

Table 10.1 Minimum per-capita acreage requirements for a quality environment

It should be emphasized that this estimate of 5 acres (2 hectares) per person applies to a self-sustaining region with good soils, a temperate climate, and abundant rainfall; requirements would be larger in areas with a less favorable climate. Since Georgians now enjoy 10 acres per person, we conclude that optimum population density (again on a self-sustaining basis at an American level of affluence) is no more than double the present population.

In this model (Table 10.1), about two-fifths of the total requirement is designated as natural environment. When we consider that food and fiber-producing lands contain considerable natural elements that contribute to life support and recycling, this estimate comes close to the 50% figure previously suggested as a working hypothesis for planners.

Appendix

To illustrate how the model of Figure 10.1 could be used, hypothetical data based on reasonable expectations for energy flows and exchange coefficients were fed into an analog computer and the output plotted as a performance curve, as shown in Figure 10.2. The "sum of values" on the Y-axis is the sum of the three values shown in Figure 10.2, and is plotted as a percentage of the maximum sum. In this highly generalized run, the optimum plateau covers a broad range between one-third and two-thirds developed lands. Anything more than 60% developed (or anything less than 40% natural) environment resulted in a precipitous decline in the value of the total environment. It should be emphasized that the optimum mix between developed and natural

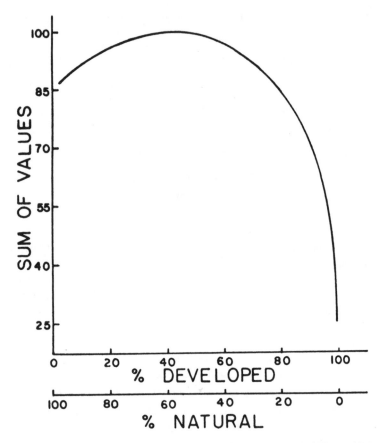

Figure 10.2   Performance curve generated from the model of Figure 10.1 using hypothetical data based on reasonable values for energy flow in highly developed and natural environments. The "sum of values" is the sum of three values shown in Figure 10.1 plotted as a percentage of the maximum. W. Smith, H. McKellar, and C. Littlejohn of the University of Florida obtained the diagram using an analog computer using light settings of percentage of developed land.

environment could vary considerably from region to region depending on the intensity of development, the kind and amount of poisonous wastes produced, the capacity (productivity) of the natural environment, the density and behavior of the human population, and so on. However, performance curves of the type shown in Figure 10.2, appear to be characterized by rapid decline — once one goes beyond the optimum plateau. If this is a true generalization, it may explain why it is difficult to recognize overdevelopment before it is too late.

During the coming year, it is hoped that data from actual situations can·be used to further test and refine the procedure. Particularly desired are data on actual economic budgets, power flows, and land-use patterns in large metropolitan districts, or other regions or sections that have a functional unity. First, attention will certainly be given to areas where research and planning inventories can provide accurate values, and where public opinion and government organization are sufficiently strong to promote serious planning aimed at preventing overdevelopment. In addition to showing energy relationships, the energy diagram is a way of writing differential equations, and the differential equations are an intermediate step in putting the model on an analog or digital computer. The translation of the Figure 10.1 model is given as three linked equations below in which the natural energies are designated S, fossil fuels F, the developed lands D, the natural lands $(1 - D)$, the natural ecosystems N, the developed ecosystems and cities C, and the interactions of diversity of man and nature I; k are the coefficients for each pathway that may be evaluated from data. Where the pathway has little drain action on its source, it is indicated by a small triangle and its outflow action omitted in the equations:

$$N = k_1 \, S \, (1 - D) - k_2 N - k_3 CN,$$

$$C = k_4 \, DFN - k_5 CN,$$

$$I = k_6 \, CN - k_7 I.$$

Value rates (V) are the sum of the energies of replacement and maintenance and are thus the sum of three rates of energy flow:

$$V = k_2 N + k_5 C + k_7 I.$$

The graph in Figure 10.2 is the steady-state sum of value flows found with one set of coefficients (k) in the set of equations on an AD-30 analog computer.

Energy, Ecosystem Development, and Environmental Risk

Eugene P. Odum

*P*ublic concerns about environment, fuel shortages, pollution, crime, and other social disorders are, as viewed by the ecologist, symptoms of the basic developmental problem that has plagued society on a micro or local scale in the past, but which now must be coped with on a macro scale, that is, on the state, national, and global levels. Simply stated, the problem is how to make an orderly transition from pioneer rapid-growth civilizations to mature civilizations, as is dictated by the natural limitations of the earth, and the rapidly rising expectations of even larger numbers of people. In terms of the principles of cybernetics (the science of controls), the problem is how to move from the positive transient state, where positive feedback promotes growth, to the steady-state or to at least a low-growth state, where negative feedback promotes balanced inputs and outputs, and guards against excesses that could destroy the system.

The term "feedback" refers to that part of the output (energy or money, for example) "fed back" into the system to accelerate (positive) or decelerate (negative) a rate function. Positive feedback that accelerates growth is necessary and desirable in a society that has not yet fully utilized its available

Originally published in *The Journal of Risk and Insurance* (formerly The Journal of Insurance), **43**(1), 1976, a publication of The American Risk and Insurance Association.

resources, but to maintain control, energy, and money have to be diverted from growth to maintenance. The process is painful in a country such as the United States, where growth is nearly an "ethic" or "cult" and exciting, as opposed to maintenance, which is considered dull and uninteresting.

Economic and Demographic Transition

The greatest of all challenges is how to make the transition from growth to maturity so as to avoid the third possible state: the negative transit state that leads to aging and death. Individuals experience all three states: youth, maturity, and aging. However, it is important to emphasize that large complex systems on the level of ecological ones, such as an ocean or a city, do not necessarily have to age and die after cessation of growth and achievement of a mature state (for more discussion along these lines, see Odum, 1969). Individuals come and go, species evolve and become extinct, societies rise and fall, but, at least up to now, large natural systems and civilizations have continued. No theoretical reason exists why a well-ordered system of man and nature cannot maintain an equilibrium state and continue to improve the quality of human life.

To accomplish the transition from youth to maturity, components must be reordered and excessive growth stimulants have to be removed, such as growth-promoting forms of taxation, deficit spending, and government subsidies that promote non-self-sustaining growth. Therefore, economic and social systems appropriate for growth systems but inappropriate for the saturated level must be replaced by appropriate components if the whole is to survive. Already seen is a sort of natural death of certain growth systems, for example, "colonial capitalism" in which nations grow rich by exploiting the resources and people of poor nations, with little or no return benefits.

The negative feedback exerted by OPEC countries is coming almost too fast for the economic and political worlds to cope with it, but it cannot be ignored. This observation does not mean that capitalism as a basic economic procedure is outmoded; its flexibility and capacity to evolve from a strictly capital-intensive system to a more human-intensive system is being tested. The point is that no system of man or nature survives for long unless it is capable of orderly changes that are adaptive in the face of competition from alternative systems.

The Entropy Law

Because society has generally proved inept at anticipatory planning, but has often demonstrated skill in solving crises when they appear, the general problem can be stated in crisis terms as follows: The great challenge is how to use available energy and ingenuity to reduce to tolerable levels the disorder or entropy that is an inevitable consequence of growth in size and power density. As systems grow larger, more complicated, and require larger transformation of energy from one form to another, disrupting or disordering forces increase and must be handled (the entropy of the second law of thermodynamics).

In terms of the MIT-sponsored study *The Limits of Growth* (Meadows et al., 1972), the challenge is how to avoid the worldwide boom and bust that is likely if attitude and management strategies are not altered to establish reasonable negative feedback control, especially with regard to human population size, the management of energy procurement, and the technology of resource conservation and reuse. The authors of the *Limits of Growth* study have modified their original oversimplified models to show how negative feedback and new technology possibly can prevent the boom and bust, or at least soften its impact (Meadows and Meadows, 1973).

Spreading the Risk

The insurance industry, as with many others, seems to thrive on growth, but it also is a major maintenance industry because it is a device organized by society to reduce disorder. By spreading the risk over a wide territory and over a variety of plans and by reducing the risk whenever possible, insurance reduces the disordering effect of local or temporary perturbations, thus enhancing the stability of the whole. Nature, through millions of years of evolution, has evolved a natural selection strategy that not only works for the survival of species, but also by spreading the risk reduces the possibility of extinction of groups of species that play a common role in the ecosystem (and therefore are fundamentally indispensable). Diversification plays a key role in nature's plan as it does in insurance.

Underlying this strategy is the basic conflict between specialization and diversification. The latter is the safer strategy, but requires diversion of energy from growth, thus lowering output. On the other hand, specialization or monoculture (i.e., putting all one's eggs in one basket) increases output from exploitable resources, but also increases the risk of boom and bust to extinction. These strategies should not be viewed as either/or propositions but as a problem of achieving a rational coexistence. Evolutionary biologists have become interested in detailed analysis of how nature spreads the risk (den Boer, 1968; Reddingius and den Boer 1970).

As the ecologist views it, the insurance industry, and especially that segment concerned with evaluating risk, will become an increasingly important regulator in human affairs as society is forced (like it or not) to expend more energy and effort: (1) to pump out the disorder inherent in large high-powered systems (or, to phrase it another way, to suck up order from the environment to maintain such systems), (2) to preserve and increase the quality of life, and (3) to plan for controlled growth and the equilibrium state.

In terms of the cybernetic analogy, the insurance industry can exert powerful negative feedback for the good of society by refusing to under write high-risk undertakings and developments that clearly have a negative long-term cost–benefit. Expensive structures on floodplains or unstable shorelines and untested atomic power plants are two examples. Thus, the insurance industry can no longer sit back and calculate the risk on the basis of conventional random statistics, because disasters increasingly are no longer random events, but are man-made, often deliberately promoted by vested interests and encouraged by government subsidy for short-term economic gain. Documentation of an environmental example of this trend appears at the end of this paper. The hope is that many will agree that the insurance profession should take a vigorous stand against government ensuring developments that the industry cannot ensure because of their risky nature. For the government to ensure them would amount to economic bailouts, thus supporting bad business.

Every aspect of human endeavor now faces the problem of dealing with short-term crises in terms of long-term consequences. As noted, society is experienced and skilled in short-term manipulations, but it has few historical, economic, or political precedents or guidelines for long-term technological assessments and for comprehensive planning; these are indeed new ventures. Thus, society must have the patience to proceed in orderly fashion from: (1) public education to create awareness of basic problems, (2) to analyze options available, and then (3) to take the necessary political action. Attempts to legislate solutions without first allowing time for the first two processes to run a reasonable course most likely will do more harm than good. As R.G. Ridker, an economist for the private organization Resources for the Future, has so aptly expressed it, there must be:

time for vestiges of attitudes and institutions developed in frontier days to die
out, time for power struggles between vested interests groups to be played
out, time to devise and to implement solutions. (Ridker, 1972)

To establish rational policies for these new ventures, the traditional reductionist approach of science and technology must be coupled with a holistic
approach based on properties that are unique to the integrated system of man
and nature. For this reason, the basic concept of the ecological system or
ecosystem should be reviewed with emphasis on the key role played by
energy, the only real common denominator to the work of human beings and
the work of nature. Also important are specific examples of recent advances
in ecosystem science that show how the risk of disorder can be greatly
reduced by designing with rather than against natural forces in efforts to get
what is needed and wanted. One major theme is that major disasters resulting
from storms, floods, droughts, and the like are increasingly man-made and
not natural.

The Ecosystem Concept

Not too long ago, a presidential message to Congress contained the following
statement: "We have begun to see that our destinies are not many and
separate at all — that is, in fact, they are indivisibly one." Such a holistic
viewpoint underlies the longstanding ecological concept of the ecological
system or ecosystem. In broad, nontechnical terms, an ecosystem consists of
living components (organisms, including humans) and nonliving components (the physical environment) functioning together as a whole according
to well-defined natural laws. The fact that human beings alter nature and
introduce new power flows (chiefly fossil fuels at the present time) to
supplement the energy input of the sun does not in any way change this basic
concept since people-developed ecosystems, such as cities, also function as
wholes according to the same thermodynamic and other natural laws. Economic systems also must function within the constraints of these laws, as
economist Georgescu-Regan so well pointed out in his book *The Entropy
Law and the Economic Process* (1971).
    Fundamental to the holistic concept is the hierarchical nature of the
systems of humans and nature. As components are added together to create
larger functional units, additional attributes come into focus — attributes not

present, or not evident, from the behavior of components functioning separately. When hydrogen is combined with oxygen in a certain manner, for example, water is formed, an entity with entirely different properties from either of its components. Likewise, when trees evolve together they form forests with an entirely new set of attributes; in other words, the forest is more than just a collection of trees. Similarly, humans and nature form a coupled system with properties of the whole that can neither be understood nor managed as long as the components are treated as separate entities (for additional discussion of ecosystem concepts, see Odum, 1975).

   Although the concept of the ecosystem, and the general idea of the oneness of humans and nature, have been widely accepted as a theory, society has yet to put such theory into practice. As long as the supply of resources and environment exceeded demand, society has not deemed it necessary to practice ecosystem management because problems could be solved one at a time as they appeared. Thus, nearly all applied natural science (i.e., technology), political science, economics, and law in the past have dealt with components rather than with the whole. What happens to a forest, for example, has largely been determined by what competing self-interest (for example, paper interest, lumber interest, real estate interest, water interest, recreation interest) gains the economic and legal upper hand — often with little regard for considerations of total worth and the future welfare of society. The same observation can be made for the present management of cities.

   The environmental awakening is clearly the result of realization by larger and larger numbers of people that society must begin to study, analyze, and act in terms of larger and larger units — that is, whole cities, states, regions, nations, and regional groups of nations. Ultimately and logically, this reasoning means working rapidly towards international coordination of management strategies and rules of conduct, as the well-being of the whole biosphere soon will be the overriding concern of all mankind. Clearly, the continued maximization of only each component as a separate entity means the optimum for the whole will be exceeded, at which point the components themselves will be inevitably degraded or destroyed. True ecosystem management, that is, decision-making based on optimizing the whole, will take time to organize, as previously pointed out. For now the groundwork can be laid and an effort can be made to avoid unnecessary disorder that could sap the energy and the will to make the transition.

Land-Use Planning

One area where the groundwork can be laid now is in land-use planning, which gradually will replace zoning as society's choice for a workable political and legal procedure to maintain the quality of the total environment. Although an oversimplification, the point can be made by contrasting two kinds of environment, which together determine the nature and quality of the total environment, and which must soon become subject to a common economic value assessment. These two kinds of environment can be labeled: (1) natural environment and (2) developed environment. Natural environment refers to any area of the biosphere essentially self-maintaining in that little or no energy or money input from the power flows controlled by people are required for continuous and orderly operation. In terms of energy, this environment is occupied by natural solar-powered ecosystems that provide the basic life support for civilization. A large natural forest, the sea, or a river are examples.

A developed environment is structured and maintained by large man-controlled power inputs that increase with the intensity of development. In terms of energy, this environment is the habitat of fuel-powered or fuel-assisted ecosystems that run on concentrated energy stored in the earth's crust. A city is the ultimate example of a highly developed fuel-powered ecosystem, but a golf course, a suburban development, or a cornfield also are developed environments because large amounts of fuel and money are required to maintain them in the developed state, even though they contain certain natural elements such as trees and grass.

Natural ecosystems differ from developed ecosystems not only in source and level of energy, but also in the manner in which their growth is controlled. For the most part, natural ecosystems have evolved effective internal negative feedback mechanisms that slow down and halt growth when the optimum is reached. Usually, although by no means always, cancerous overgrowth is avoided under nature's management. In contrast, people-developed systems have a strong built-in positive feedback that favors overdevelopment because: (1) control is largely through external power forcing functions (rather than by way of internal negative feedback functions), and (2) components with a high economic value are maximized at the expense of other necessary but undervalued components. Fortunately, ecologists and economists are beginning to establish a dialogue on closing this gap in value assessment. In the meantime, it is important to recognize the internal and external forces that tend to frustrate attempts to plan for a rational use of the total environment.

The Tendency to Overshoot

Barkley and Seckler, in a discussion of *Economic Growth and Environmental Decay* (1972), list factors which tend to force urban development to overshoot the optimum: (1) detrimental self-crowding effects such as noise, other pollution, rising costs of schools, sewers, police protection, and so on are not felt until sometime after the optimum density has been exceeded; (2) money is to be made in continued growth beyond the optimum in the absence of effective zoning and land-use planning; (3) political power is concentrated within a few wealthy groups; and (4) the mystic of growth persists from pioneer days when rapid growth was necessary and desirable. Nearly everyone agrees that the best total environment for people consists of a balanced mixture of natural and developed environments (cities, farms, industrial parks intermixed with natural forests, lakes, rivers, etc.). Yet the lack of legal and economic constraints on the growth of developed systems ensures that developed environment will grow at the expense of natural environment to the point where the quality of both will decline.

To conclude that inherent tendencies in present legal and economic dynamics tend to cause cities and other developments to grow beyond their optimal size is not to embrace an anti-human or anti-urban philosophy. Because cities are valuable, providing a source of economic wealth, they must be protected with care. Cities can be preserved only if the built-in self-destructive positive feedback is recognized and counteracted. While the desperate predicament that New York and other large cities are suffering generally is blamed on poor fiscal management, one suspects also that the large cities have overshot, not only their financial support base, but also their life-support base, as part of the financial stress arises from inability to pay for maintenance work.

Outside Forcing Functions

South Florida provides an example of how unrestrained, self-destructive development can be forced upon the region by economic forces from outside the region. Florida land has been touted as being so valuable that large investors, such as insurance companies and Swiss banks, are buying huge amounts of acreage in the expectation of reaping large speculative investment profits. At the same time, immigrants from other states, expecting quick profits in the boom, are moving in at an unmanageable annual population growth rate of 4% (doubling in 17 years). In this forced growth syndrome, land, including sand scrub and swamps, is valued in the market far above its intrinsic worth. Once large sums have been invested, unbelievable pressure is placed on local, state, and federal governments to develop expensive drainage, waste disposal, school, and highway systems so that investors may realize a profit on their investments.

The unfortunate feature of this situation is that increasing numbers of Florida citizens of modest means do not want such cancerous overdevelopment, and are now seeking legal and political power to halt or slow growth. Otherwise, economic power from without can force a rate of growth that resident citizens of Florida neither want nor can cope with. The insurance industry, as already noted, can be of assistance in the evolution of an optimum growth policy by: (1) not investing in speculative land development, (2) insuring owners of only solid residential and industrial development that includes adequate provisions for water, waste disposal, open space, and future maintenance, and (3) not underwriting housing and related construction where objective analysis shows that the area is saturated in terms of environment and services.

When there is strong opposition to a development by bona fide citizen organizations, investment interests should heed this negative feedback signal and seek other ways to use their resources to produce a net gain for society as well as for business.

Energy, the Common Denominator

Because energy consumption and general measures of economic well-being (such as GNP) show a linear relationship (Hafele, 1974, where Fig. 1 is a chart showing the linear relationship between energy use and gross national product for many countries of the world), costs can be expressed in terms of energy as well as dollars. In this manner, the work of nature and of people can be considered in the same general dimension. Thus, options on the basis of energy costs now need to be assessed more than ever. Cost–benefit analysis, which is a form of risk assessment based only on money, can be too easily manipulated to hide costs and risks such as pollution or diminution of environmental life-support capacity — aspects that can be better evaluated in terms of energy than in terms of money. We especially need to determine the energy cost of producing and using energy. For example, if a proposed atomic power plant requires two BTUs of fossil fuel for every BTU of atomic power produced, then a loss rather than a gain will be incurred over the long run, even though the plant might temporarily solve a power crisis. In an issue of *Business Week*, energy cost-accounting is headlined as "The New Math for Figuring Energy Costs" (1974). Feedlot beef is an example of an agricultural system that is bad business in terms of energy cost-accounting, even though it has been lucrative for certain agro-business operators and also a short-term tax-shelter for the wealthy. An assessment (see Fig. 5 in Steinhart and Steinhart, 1974) indicates that about ten units of fuel energy are required for every one unit of food energy produced. The United States can hardly expect to export grain or support a world grain bank for hungry nations and also waste food energy in such an inefficient ecologically unsound conversion system. The cow is naturally adapted to converting grass and other energy-cheap materials into high-protein food because of the rumen, where special microorganisms do the work of conversion. To feed cows energy-expensive grain, thus bypassing the adaptation, makes neither ecological nor economic sense. One can almost predict that this system will indeed be a boom-and-bust operation. (Insurers should not insure owners of stock in feedlots!) If so, the government will be pressured into bailing out with tax money another inefficient business because the risks were not recognized by the agricultural specialists who promoted the system. This example is just one of many that could be cited to show that ecological and economic common sense are directly related. The two words come from the same Greek root, "oikos," meaning "house." Ecology is the study of the house (people's total environment), and economics literally is the "management of the house," which must include all of the environment.

As discussed at the beginning of this section, the concept of net energy is an important one relating to energy cost-accounting. The potential or gross energy in the atom or under the sea bed is impressive, seemingly limitless, but the question is how much net energy will there be when all costs of procurement, conversion to useful work, and abatement of pollution disorder are paid. Options must be considered in terms of what energy sources or combination of sources have the best probability of the highest net yield. Decisions should not be made solely on the basis of technological feasibility. Net energy options are the ones that should be insured and on which government research and development money should be spent rather than those with a high risk of negative yield.

To place this discussion in broader perspective, the fundamental interrelations of energy, people, and nature can be summarized as follows:

1. In a technologically advanced industrial society, energy itself is not likely to be limiting, but the consequences of converting energy from one form to another are limiting. When fossil fuel is exhausted, plenty of poten-

tial energy is found in the atom and the sun, but tapping such new sources will be more difficult (i.e., the technology will have to be more sophisticated) than the relatively simple procedures of burning coal and oil. Accordingly, it will be a real challenge to keep the unit cost of a kilowatt or BTU from rising. In fact, the probability is that unit costs will rise despite all best efforts. Also, heavy retooling or transition economic losses always are incurred when it is necessary to shift from one technology to another.

2. No technological way exists to bypass the second law of thermodynamics. One can ameliorate thermodynamic disorder created by energy conversion, but one cannot avoid the basic cost of dealing with it. Concentrating and intensifying the use of energy within small areas (cities, heavy industries, etc.) creates especially difficult thermodynamic disorder problems.

3. Exponential growth (such as recently seen in Japanese industrial growth) cannot continue for long. Two options are available: (1) let positive feedback run its course until overshoot occurs (boom and bust), or (2) install some negative feedback to bend growth curves in order to prevent overdevelopment and thereby buy time to solve problems. The former has an inherent risk that the overshoot might be followed by a worldwide depression so deep that recovery could not be managed because all reserves had been used fueling the boom.

Lost in the controversy between the extremes of zero growth and laissez faire growth is the simple common-sense truth that a controlled moderate rate of growth not only prolongs the time when the undeniable benefits of growth can be enjoyed, but also reduces the cultural and technological lags inherent in rapid growth. If care is taken of the consequence of growth as it occurs (not later), then the risk of the overshoot is reduced (Ridker, 1973).

4. As the size of a system increases, the cost of maintenance of a network of services increases at least as a square of some kind of power function:

$$C = \frac{N(N-1)}{2}, \quad \text{which approximates} \quad \frac{N^2}{2},$$

where N is the number of units in a network and C is the maintenance cost. Thus, doubling the size of a city or a power plant is likely to more than double the cost of maintenance. As other tradeoff advantages are available to an increase in size (i.e., economies of scale factors), bigger is better only up to a point. Theoretically, an optimum power concentration per unit area exists beyond which any increase in size costs more than it is worth. Cost–benefit curves are humpbacked, and not straight lines as often projected by the superficial optimists.

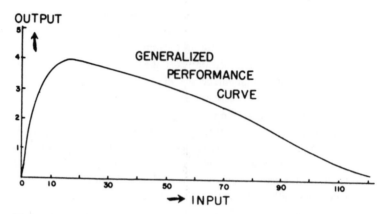

Figure 11.1   Theoretical performance curve (after Miller, 1965). This gen-eralized information input–output performance curve can be applied to several levels of biological organization. Presumably, some such curve applies to a power-in/benefit-out relationship. One insidious feature of the curve is that the downward trend that comes after the optimum is reached is so much more gradual than the pre-optimum upward trend. Thus, increases in output benefits are recognized quickly in the early stages of development, but declining benefits are not so easily recognized after the optimum has been passed. Unfortunately, as noted, large, power-hungry, people-made systems have an inherent tendency to overdevelop, as money can be made by exceeding the optimum so long as payment for the increased cost of maintenance is avoided or delayed. Locating the optimum plateau in the energy–performance curve appears to be the greatest challenge in systems research today. Energy–performance curves could be important guides for insurance as risks increase rapidly for proposed expansion to the left of the optimum plateau.

The Life-Support System

The natural environment provides the life-support system for both people and their energy-consuming machines and is the major means for tertiary waste treatment. The work of nature in pollution abatement has been undere-valuated, because, as already noted, this service is considered to be a free externality. Present cost–benefit procedures (see Figure 11.1) generally fail to include this basic dependence on natural environment until after the overshoot, i.e., until the environment is badly polluted and suddenly treat-ment that previously was free has a cost. One problem in recognizing and accounting for nature's work is that natural systems operate on dilute sun

	Power requirements (kcal/m²/year)
1. Natural solar-powered, unsubsidized ecosystems[a] (these are the basic autotrophic, self-maintaining life-support systems	1,000–10,000 (av. 2,000)
2. Solar-powered ecosystems subsidized by free natural[b] inputs of energy	10,000–40,000 (av. 20,000)
3. Solar-powered ecosystems subsidized by man-controlled, but expensive, inputs[c] of energy and materials (these are the basic food-producing systems)	10,000–40,000 (av. 20,000)
4. Fuel-powered urban-industrial systems (these are the wealth-, and pollution-, generating systems that are parasitic on ecosystem types 1 and 2	1,000,000–3,000,000 (av. 2,000,000)

[a]Oceans, natural forests, and grasslands are examples.

[b]An estuary is an example of a natural ecosystem subsidized (and its productivity thereby enhanced) by natural tidal energy.

[c]Industrial agriculture is an example of a solar-powered ecosystem heavily subsidized by fossil or other fuel at appreciable cost to man.

Table 11.1   Ecosystems classified according to source and level of energy

energy and, therefore, are low-powered compared to cities subsidized by huge fuel imports.

Table 11.1 shows the order-of-magnitude energy flows in the three basic kinds of ecosystems that must function together in some kind of balance if society is to prosper. This classification is an extension of the dichotomy discussed earlier in that developed environment is subdivided into agricultural and urban-industrial components. Cities and industrial complexes fed with concentrated fuels operate at a power level three orders of magnitude higher than those natural systems evolved to operate on dilute sun energy alone. However, nature is efficient at its level of power when one considers that natural systems are self-maintaining and generally take care of their own wastes — in contrast to cities which, as now managed, are almost completely parasitic on types 1 and 2 systems, as listed in Table 11.1.

Because of the low unit-area capacity, large areas of natural environment are required to absorb the thermodynamic disorder produced by the high-energy systems of people. Thus, what might appear to the average citizen (or the developer) as unused, empty lands and waters around the centers of human activity are in reality hard-working solar-powered systems vital to the maintenance of urban hot spots. The largest cities and industrial centers of the world are not located accidentally on large bodies of water that provide free services. When such environment is not available, or is so severely stressed that it is unable to maintain itself and also work extra for pollution abatement, then artificial tertiary treatment of large volumes of low-level wastes (CO_2, heat, NO_2, SO_2, radioactive substances, and so on) would become necessary. In the long run, these low-level/large-volume wastes, together with the outright poisons (lead, mercury, DDT, industrial phenols, and so on) pose the greatest threat to the quality of human life. If all these wastes have to be treated artificially, costs would skyrocket to the point that no known economic system could cope with them.

Based on two different models (Odum and Odum, 1972), it is suggested that in a state or region with one or more large industrialized city, at least 50% of the land and shallow waters should remain in a natural state if the quality of both city and country is to be maintained. Whatever proves to be an optimum mix of developed and undeveloped environment, one observation is certain: The developed soon will overshoot the undeveloped unless comprehensive regional land-use and energy management plans are adopted. The other strategy that should be coupled with prudent planning is one directed toward energy conservation, including seeking less wasteful and less energy-intensive technology (put the cow back on grass, as it were; or, in the broader context, design with rather than against nature).

The Sea-Beach: A Specific Example

Economic losses from storms and flooding on seacoasts and river valleys are increasing at an alarming rate, not because natural storms or floods are any worse, on the average, than they have been in the past, but because increasingly expensive and densely packed developments are promoted on unstable terrain that is highly vulnerable to just ordinary reoccurring natural forces. Sea beaches, in particular, now have such an inflated real estate value that developers will take nearly any risk provided they can purchase insurance and count on the federal government to use taxpayer money to: (1) build expensive sea walls and other mechanical devices to protect their structures, and (2) provide instant disaster relief when the devices fail, as they do more often than not. This absurd situation is the product of the notorious pork-barrel politics that some people label as "welfare projects for the rich." One way to cut federal spending is to eliminate this subsidy, but few politicians have yet to suggest such a move.

As recently as ten years ago, neither scientists, engineers, nor the public fully understood the nature of the sea beach ecosystem. Scientific study was so fragmented into separate disciplines that conflicting theories were more numerous than unifying ones. The public assumed that applied scientists (engineers, for example) knew what they were doing in constructing devices to restrain the energy of the sea, when in reality they were experimenting in the hope of finding solutions. These experiments have helped in formulating unified principles. If society can profit from the experiments and stop repeating the unsuccessful expensive ones, then money spent in the past will yield dividends in energy (i.e., money) saved in the future.

Without being too technical, an attempt is made here to review the current understanding of the dynamics of sea beaches. The author is especially interested in this subject, as his group at the Marine Institute of the University of Georgia is considered to have made significant contributions to overall theory and practice.

First, a beach is a dynamic system, constantly changing as a result of both local forces (natural and people-generated) and global changes, such as changes in sea level. Beaches experience, often in rapid succession, the three cybernetic states previously mentioned, that is, enlarging, contracting, or maintaining of steady-state. If one looks at a large area, such as the Atlantic or Gulf Coast, the acreage of good beach has, until recently, remained about the same as the area that has been subject to erosion. Now, however, more is lost than gained, primarily due to society's inappropriate or poor management.

The key to predicting what a specific stretch of beach is doing is the input–output sand budget, just as the key to the status of a business is its financial balance sheet. When the sand budget is balanced — that is, as much sand is deposited on the beach as is removed by the tides and waves — then winds build up sand dunes on which vegetation develops in an orderly sequence as follows. First come the hardy beach grasses, such as the northern Ammophila and the more southern sea oats. Then there are the herbs and vines, such as beach morning glories, and salt and wind-resistant shrubs. As the sand is held by the vegetation and dunes accumulate to higher levels, junipers (cedars), pines, and oaks enter, creating attractive dune forests.

In planning recreational and other development, it is recognized that, insofar as possible, dunes and their vegetation should be preserved. What has not been fully recognized is that a shift in the sand budget may place the dune in an entirely different role in relation to the beach strand. If the sand budget becomes negative, as a result of changes in offshore currents, perhaps caused by a severe storm or by dredging, then the beach may begin to shift landward and the dunes will begin to erode despite vegetative cover. The dunes then become a source of sand to renourish the beach. In time, if the natural events could be left to run their course, the beach likely would stabilize in a new position, with the process of dune building repeating itself.

Before this interaction of geophysical and biological forces was understood, building expensive sea walls, groins, or other barriers was thought to be the answer to beach erosion problems. In many cases, this approach has proved not only futile, but also it may have hastened the erosion of the beach for two reasons:

1. With a sea wall the wave forces are directed back on the beach instead of being dissipated gradually over large expanses of sand and dunes, resulting in a steeper beach, and the full force of storms batters down the walls, requiring frequent and expensive maintenance. At Miami Beach, for example, no longer is any beach exposed at high tide, and a constant undermining of the sea wall occurs. One must wait for the tide to go out to enjoy a beach.

2. A second, but more subtle, effect of barriers is that the sand supply stored in the dunes is cut off by the obstruction and is unavailable for natural renourishment.

Engineers have experimented with an even more expensive scenario to overcome the failure of the barriers, namely, sand-pumping or artificial sand nourishment. At a meeting on beaches it was revealed that it cost two million dollars per mile annually to sand-pump a public beach to balance a negative sand budget. Much oil is burned up in such an effort, oil desperately needed to conserve energy for more vital uses. For further information, see the well-illustrated article in *American Scientist*, and for a more technical ac-

count of the sand budget concept see Technical Report 78-2 of the Georgia
Marine Science Center (Dolan et al., 1973; Oertel, 1973).

The prudent procedure is to recognize the inherent instability of low-lying
shores and to regulate the development of people-made structures accordingly. Where shores are in the public domain, this can easily be done (once
the "pork barrel" is declared obsolete) as recreational or other facilities can
be shifted to take advantage of stable or growing strands. The problem is
more acute in private ownership, especially where there have been large
capital investments that cannot be readily moved. The person who owns
beach property that is accreting (i.e., growing) stands to profit, while the
person stuck with eroding property calls for government subsidies to protect
the property. On one short stretch of the Georgia coast, one man's beach
property has increased from 30 acres when surveyed in 1900 to 130 acres
when next surveyed in 1970. He has been offered two million dollars for the
newly created 100 acres on which he has paid no taxes or otherwise spent
any money on maintenance during these years of accretion. Whether real
estate created by the ocean is to continue as private property or become a part
of the public domain is a question for the courts to decide, but as of now the
law allows him to profit from the free work of nature. However, just a few
hundred yards along the coast, other owners invested in expensive houses
that are, one by one, falling into the sea as that part of the beach erodes.

New Zealand, a country with a long shoreline, has developed a national
policy that is at least a partial solution to the conflict between public and
private interest. Owners of seaside property in that country retain full private
property rights, including rights to exclude the public, as long as they use
their property for their own enjoyment or personal business. If they decide
to profit by developing the property for intensive use, then they must deed
the seaside strip to the state. The beach then can be enjoyed by the public,
and no structures can be built on it. In this way, building structures are located
away from the shore, where risk of storm damage is less and where the
sweeping view of the water, beach, or rocky shores is available for both
private property owners and visitors who now have access to the water's
edge. Many states and counties in the United States are adopting beach
protection or shore protection acts that restrict future development to stable
terrain away from the strand and the dunes. Perhaps, soon, a national policy
will evolve.

Much of what has been said about sea beaches also applies to flood plains
and inland wetlands. The 1973 Mississippi flood was the most costly in the
history of the United States, even though in terms of the amount of rain that
fell in the watersheds it was an ordinary natural event, not so extreme as

many previous spring floods. The combination of diking to restrict the flow within narrow channels and the increased rate of runoff from the concrete of cities and highways and channelized (i.e., ditched to make water move downstream faster) streams caused water to rise increasingly higher in the straight-jacketed river. When water overflows one of the high walls, a destructive torrent is released, wiping out everything in its path and forcing water to rise even higher than the natural floodplain.

No such destructive flood could occur in a natural floodplain, as water would not run off so fast and would spread gradually over large expanses of floodplain, where it would fertilize the forests and enrich bottomland crop fields and not kill anyone because plenty of time would be available to move out temporarily. Paradoxically, the people-made floods are often followed by severe crop failures on the uplands because the water rushes to the sea instead of sinking into the water table, where it is needed in times of deficient rainfall. The 1973 Mississippi flood conditions have been well analyzed in an article by Belt (1975). The last sentence in his paper reads as follows: "The 1973 flood's record was man-made."

Conclusions

In summary, the theme of this paper has been that society is unnecessarily increasing environmental risk and that so-called "natural disasters" are increasingly people-made. As long as people and the insurance industry write off these disasters as natural and therefore unavoidable, then nothing will be done except the encouragement of a frantic and futile effort for a more expensive and energy-consuming confrontation of powerful natural forces. If, on the other hand, human beings accept part of the blame, then certainly they cannot only correct their mistakes but can in many cases direct natural forces to their advantage instead of amplifying the harmful impact. Flood control, for example, is a matter of managing the entire river valley system, not just building restraining devices along the bank.

Most of all, the insurance profession is urged to take an active part in the redesign process by instigating a broader cost–benefit analysis for evaluating risk by including energy cost-accounting and by becoming more vigilant in opposing both private and governmental vested interests, which would sacrifice the future for a short-term gain or for a quick-fix solution. Best of all, the insurance industry can strike a significant blow for energy conservation, as unnecessary confrontation with the forces of nature wastes energy.

The Transition from Youth to Maturity in Nature and Society

Eugene P. Odum

*I*n teaching, we find that the best way to impart new knowledge and new or expanded ways of thinking is to relate the unfamiliar to the familiar. That is, we start with something a person already knows or has experienced and go on from there. Sometimes drawing analogies helps. Accordingly, in presenting my "transition" theme, I shall start by drawing parallels between individual growth, which we all experience, and the development of natural communities and human societies.

Individual development involves going from youth to adulthood. In a somewhat parallel way, human societies go from pioneer stages to mature stages. In the individual, the transition is called "adolescence," while at the societal level the transition has been called the "demographic transition." In both cases, the transition is painful and critical in determining what maturity will be like. A major difference is that the transition in the individual is programmed by the genetic system, so we go from childhood to adulthood at a certain age whether we like it or not. Growth hormones are turned off; maintenance hormones are activated. In contrast, societies as well as natural communities mature as a result of negative feedback with no set-point as to when the transition might occur. Often, changes are gradual over long

---

This essay is based on a Graduate Commencement Address at the University of Georgia in 1988 and published in *Bridges* 2(1/2):57–62, 1990.

periods of time. There is much controversy as to whether the demographic transition will occur "naturally" or "laissez-faire", or whether because of the momentum and danger of overshoot it is desirable to speed up, or "manage," the transition by political and/or economic means.

An important point to be made is that at both levels (individual and community) there are many processes that are quite appropriate and necessary for survival during youth but quite inappropriate and detrimental to health in maturity. For example, rapid growth of tissues, organs, and body size is necessary and expected during childhood, but such growth in adulthood is very likely to be cancerous. Likewise, high birth rates, large families, and exploitation of unused resources are important for survival and development of pioneer communities or countries, but all of these things become counterproductive when population density becomes great, space and resources become limiting, and the environment becomes more and more stressed by pollution.

My theme, then, is that there are striking parallels between development at these different levels of organization. Furthermore, I contend that study and understanding of the processes involved may help us overcome the resistance to change inherent in political and economic systems, and thus help us not only to survive the demographic transition but to profit from it.

Development of Natural Biotic Communities

Developmental change that occurs in natural communities is termed "ecological succession" by ecologists. It can be readily observed when an agricultural field is abandoned and left for nature to redevelop, or when a new pond is created by damming a catchment basin. During the early stages of development, opportunistic species that grow and reproduce rapidly colonize and exploit the unused nutrients, light, and other resources. As time goes on, the quantity of life or "biomass" increases and changes in species composition are rapid as more species invade and there is intense competition for resources. Gradually, the species that dominate the community are those that are competitive in the crowded resource-limited situation that is the inevitable result of any kind of rapid growth. Individuals of such species are generally longer-lived and produce fewer offspring than the pioneer species. For example, a ragweed plant — a species of the early stages of old-field succession — is about a third by weight composed of reproductive structures and numerous seeds are produced, while a toothwort plant, which lives in the shade of the mature forest, puts less than 1% of its energy into reproductive structures that produce only a few seeds. Most of the available energy goes into the production of leaves, stems, and roots needed for survival and maintenance of the individual in the light-limited environment. (Comparison of these two plants is based on a study by Newell and Tramer, 1965.)

Interaction between species also changes during ecological succession. In the transition from pioneer to mature or "climax" natural communities, what amounts to "cutthroat" competition that results in rapid replacement of species is gradually replaced by more cooperation or "mutual aid" between species, which enhances the conservation and recycling of resources and the survival of "climax" species that prosper under conditions of tight resource availability. (For more on ecological succession, see Chapter 7 in Odum, 1989a.)

The importance of mutual aid in mature communities has been especially well-documented in studies of coral reefs and tropical rain forests. For example, the coral itself is a combination plant–animal structure. Algae grow right inside the tissue of the animal "polyp" and in the calcareous skeleton. Accordingly, there is rapid and efficient exchange of food (that the algae convert from sunlight) from plant to animal and key nutrients (such as phosphorus excreted by the animal) from animal to plant. Reefs on Pacific atolls are prosperous natural "cities" packed with life despite the fact that the surrounding oceans are nutrient deserts. Recycling and retention of nutrients and conversion of sun and wind energy are so efficient that the reef is highly productive with very little resource imports. (For more on coral reef mutualism, see Muscatine and Porter, 1977.) Contrast this with a productive,human-designed crop field (which, in the ecological context, is a very youthful community) that requires huge imports of fertilizers.

Mature Amazonian tropical rain forests in South America exhibit numerous adaptations for maintaining luxurious vegetation on relatively poor sandy soil. (Rain forest adaptations are reviewed by Jordan, 1982.) Many of these adaptations involve cooperation between unrelated species for mutual benefit. For example, partnership between fungi called mycorrhizae ("fungus root") and tree roots enables trees to collect and retain scarce nutrients more efficiently than is possible by the roots alone. The tree provides food for the fungus, which in return supplies minerals needed by the tree. This fungus–root symbiosis is widespread not only in the tropics but in vegetation on poor soils in the temperate zone as well (see Wilde, 1968).

Perhaps the most basic change that occurs in the ecological transition from youth to maturity involves a shift in allocation of energy from growth to maintenance. In youthful communities, most of the primary production (i.e., photosynthesis) goes into growth, that is, increasing the size and diversity of the living system; at this stage, the biomass is small and does not require much energy to maintain. In sharp contrast, the large and complex organic structure of the mature system mandates that most of the available energy must go into maintaining and improving the quality of the infrastructure that has already been produced. And, of course, the same basic switch in the allocation of food energy occurs during adolescence in the development of the individual.

Youth and Maturity in Society

Just about all of the developmental processes and transitions we observe in nature (and also in the development of the individual) have their parallels in the development of human societies, granting that cause-and-effect relationships may be different. In pioneer societies, as in the youthful stages of the natural community development, high birth rates, opportunistic exploitation of the environment, and accumulation of resources are necessary for the survival and growth of human societies. Slavery and other exploitation of people have also been common in the early stages of societal development. As societies mature as a result of increased density, development of expensive-to-maintain urban infrastructures and increasing demands on environment and natural resources that were appropriate in youth become increasingly inappropriate. Human rights are given higher priority and growth has to slow down and be managed. Many attitudes and economic policies have to be reversed in order to avoid the typical "boom-and-bust" cycle. Humans have a historical propensity to wait too long to make the adjustment, perhaps because youth is both enjoyable and profitable!

On the other hand, to draw an analogy with individual development, it is important that "about-face" changes be made gradually and holistically. One of the main problems of growing up in today's stressful society is that children are often forced to make adult-level decisions about sex, drugs, education, etc., at too early an age, thereby making the adolescence transition very difficult. Many well-meaning environmentalists cry out for desirable changes but at an unrealistic speed, such as proposing instant recycling of everything, thereby making the demographic transition more difficult.

In summary, I have suggested that there are important lessons we can learn from the study of development in nature. For example, how to cope with the fact that energy and resources (as well as money and taxes) have to be diverted from growth in size to maintenance of the complex infrastructures (such as cities, transportation and communication networks, parks, recrea-tional facilities, and so on) that have already been built. As is readily observed, taxes to pay for services increase roughly in proportion to the density of urban development. For example, New York's per-capita state and local taxes are three times as high as Mississippi's, a much less densely populated state. (For the correlation between taxes and urbanization, see Table 3.17, p. 161, in Odum, 1983.) The politically popular policy of not raising federal taxes simply results in rising local taxes. We need to be less concerned with increases in taxes (which in the United States are far lower than in Europe) and more concerned that our tax money is used efficiently for "pumping out the disorder" (crime, decay, and so on) inherent in large mature systems.

In other words, quantitative growth must gradually give way to qualitative growth and greater concern for quality of life. In the United States, we see this transition underway already since service jobs are increasing while production jobs are decreasing, and there is greater concern for environ-mental quality and land-use planning. As noted previously, coral reefs and rain forests prosper in a world of limited resources by efficient recycling and waste reduction (see Odum, 1989b for a discussion of the importance of waste reduction) and especially by developing extensive cooperation be-tween units for mutual benefit. (Cooperation in nature and human affairs is analyzed in Axelrod, 1984.) In human society, we see the beginnings of worldwide recognition of the need to shift from military confrontation to mutual aid in the dramatic changes that are occurring in the relationships between the superpowers — the United States and the Soviet Union. All nations need the money and resources that are now wasted on confrontation in order to direct the demographic transition in such a way as to promote family planning, environmental quality, and the kind of economic develop-ment that does not involve wasting future assets. ("Wasted Assets" are documented in a report with that title published by the World Resources Institute in 1989.)

One thing seems certain: economic development and environmental pro-tection, which in the past were considered separate issues, must now be joined, so expect to hear a lot about ecological economics in the near future. (For more on the transition from youth to maturity, see the epilogue in Odum, 1989a.)

Reduced-Input Agriculture Reduces Nonpoint Pollution

Eugene P. Odum

*R*enewed interest in using legumes to provide nitrogen for cash crops is just one of many indications that the stress of nonpoint pollution on regional and global life-support systems is bringing about a major change, almost an about-face, in the philosophy and technology of management — not just for agricultural systems but for all production systems (see Figure 13.1). Contamination of surface and groundwater by agricultural chemicals, soil erosion from both urban and rural landscapes, increases in greenhouse gases, and acid rain from power plants currently pose the greatest threats to the earth's life-supporting atmosphere, soil, and water bodies. Such nonpoint pollution sources, unlike point sources, cannot be controlled from the output side of a production system; they can only be controlled from the input side by what I call input management.

The attention of agronomists and other production managers has for many years focused on increasing outputs, such as yields. Spectacular increases in yields of cash crops have been obtained partly by selection of high-yielding cultivars, but mostly by vast increases in the inputs of machine energy, fertilizers, and pesticides. For example, 50-year trends in Georgia agriculture were recently plotted as part of a study of changing land use in the state. Between 1935 and 1985, yields per acre of major cash crops increased

---

Reprinted from the *Journal of Soil and Water Conservation* **42**(6):412–414, 1987.

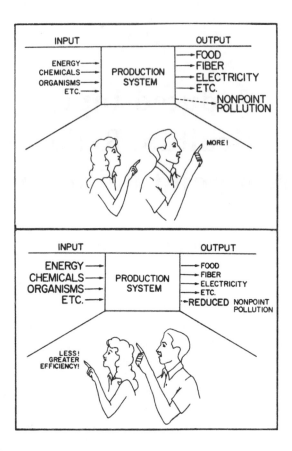

Figure 13.1 The "about-face" on management of production systems. **Top:** Focus on output, such as yield, with consequence of increased nonpoint-source pollution. **Bottom:** The shift to input management with focus on efficiency and reduction of costly and environmentally damaging inputs so as to reduce nonpoint-source pollution.

fourfold during the same time period the volume of nitrogen fertilizers used increased eleven-fold. Increased use of pesticides and herbicides shows a similar trend.

In general, crop yields have leveled off during the past ten years, indicating that diminishing returns are being reached for this kind of output management. Currently, the farmer cannot increase his or her profits by increasing yields because the cost of producing a unit of crop (bushel of corn, for example) is greater than the market price for that unit. Most important of all is the fact that increased use of agricultural chemicals in continuous monocultures is producing an unwanted increased output of nonpoint-source pollution. For example, R.A. Smith and associates (1987) reported that, while sewage pollution has declined over the past 15 years in many of the nation's rivers, nitrates, pesticides, and certain toxic metals have increased. Even more serious is the contamination of groundwater by these chemicals.

From both the ecologic and economic standpoints, it is evident that attention now must focus on the input side of the production system. When large amounts of expensive chemicals and water leach or run off, and thereby become unavailable to crops, the system can be considered inefficient and wasteful. By increasing the efficiency of such wasteful production systems, costly inputs can be reduced without too much, if any, sacrifice in yield. And the profit margin also will be increased as input costs are reduced. Accordingly, plant breeders and biotechnologists should focus on producing more efficient cultivars, not just high-yielding ones that require expensive and environmentally damaging subsidies. This, in fact, was one of the major recommendations that came out of a meeting of plant scientists to consider "Crop Productivity — Research Imperatives Revisited" (Tangley, 1986).

F.H. Buttel and four rural sociology colleagues (1987) at Cornell University have suggested that the term "reduced-input agriculture" includes "organic farming," "alternative agriculture," and "regenerative agriculture," terms that are more or less synonymous. This seems to be an excellent designation for the application of input management in agriculture because reducing both cost and nonpoint pollution is desirable.

Introducing legumes into cropping systems, either in seasonal rotation or interplanting with grain, reduces the need for nitrogen fertilizer that is bound to become more expensive with the inevitable rise in the price of petrochemicals. In our small-scale experiments a winter–spring stand of crimson clover supplied all of the nitrogen needed for a summer crop of grain sorghum or soybeans under both conventional and no-tillage cropping systems (Groffman et al., 1987). Others have reported that a small addition of mineral nitrogen may be needed for some crops, depending upon soil and water conditions (Frye et al., 1985). Organic nitrogen fixed by legumes tends to be released more slowly and retained within the crop field to a greater extent than an equal amount of nitrate fertilizer.

The increase in scale of continuous monocultures of cash crops has contributed to the increase in nonpoint pollution. Monoculture has a number of advantages; for example, it is adapted to machine culture, large-scale farming, large-scale marketing infrastructure, and aerial spraying of pesticides. Moreover, the economics of scale are favorable in the short term. On the other hand, monoculture has the following disadvantages: (a) it is hard on the land (which needs rest and a change of pace like other living systems); (b) it requires subsidies that are increasingly expensive; (c) it is vulnerable to pest outbreaks; (d) genetic diversity tends to be reduced; and, of course, (e) soil, water, and chemical runoff are high. Over the long run monoculture tends to be "boom and bust" (Adkisson et al., 1982).

Limited tillage and especially no-tillage cropping systems are excellent examples of reduced-input practices that reduce unwanted outputs. In the hilly Piedmont region of Georgia, Fred White and associates (White et al., 1981) estimated average annual soil losses under conventional tillage (plowing twice a year) at 11 tons per acre on good farmland and 15 or more tons per acre on marginal farmland. Annual nitrogen losses were estimated to be 10 and 15 pounds per acre, respectively. No tillage involving rotation of winter clover and rye (the latter to provide mulch) and summer grain can reduce these losses to a level below the soil and nitrogen regenerative capacity of good land. Accordingly, over the years soil quality is improved rather than degraded.

Gebhardt and colleagues (1985) report that 22 to 45% of cropland is now under conservation tillage management (highest in the central Great Plains, where water conservation is of utmost importance, and lowest in the Pacific Northwest). Of this, 4 to 48% of conservation tillage is in no-till.

As suggested at the beginning of this commentary, the concept of reduced-input or input management applies equally well to other production systems, such as power plants, industry, automobiles, and so on. For example, removing the sulfur or other acid-producing materials from coal or other fuel before combustion in the power plant is far more effective in reducing acid rain than trying to scrub stack gasses. The same is true in dealing with automobile exhaust and industrial toxic wastes. The bottom line, then, is this: The only way to reduce the threat of nonpoint-source pollution to global life-support systems is to manage more efficiently the inputs. Fortunately, there are increasingly strong economic reasons for doing just that.

When to Confront and When to Cooperate

Eugene P. Odum

Among the things we can learn from nature is that there are times and places both for confrontation and for cooperation. Or, in ecological terms, competition oftentimes is important in natural selection while at other times and places cooperation has greater survival value. In general, we see competition playing a major role in the early or pioneer stages of ecological succession, while cooperation seems to increase in the mature stages, as, for example, an old-growth forest. Also, cooperation between species for mutual benefit becomes important where soil or water are nutrient-poor as in Central Pacific coral reefs or in some rain forests. The coral is essentially a partnership between the animal polyp and the entozoic algae that enables the coral to retain and recycle nutrients and thereby prosper in nutrient-poor ocean waters. Partnership between tree roots and mycorrhizae fungi enable rain forests to prosper on poor, sandy soils.

Soon after Darwin, the Russian Peter Kropotkin published a book entitled *Mutual Aid: A Factor in Evolution* (1902/1935) Kropotkin chided Darwin for his overemphasis of natural selection as a bloody battle (the "red tooth and claw" metaphor). He outlined in considerable detail how survival was often enhanced or even dependent on one individual helping another or one species aiding another for mutual benefit. His writings were much influenced by his personal philosophy of peaceful coexistence. Like Gandhi and Martin

---

First published in *The Georgia Landscape,* The School of Environmental Design, University of Georgia, 1992.

173

Luther King, he was a firm believer in peaceful or nonviolent solution to human conflicts. A good deal of his book is devoted to documenting the importance of cooperation in primitive societies, rural villages, labor union guilds, and so on.

In all fairness to Darwin, we should point out that he was well aware of the existence of indirect and nonviolent natural selection. For example, in *Origin of Species* he wrote, "as the mistletoe is disseminated by birds, its existence depends on birds: and it may metaphorically be said to struggle with other fruit-bearing plants to tempt birds to devour and thus disseminate its seeds rather than those of other plants."

To me, Kropotkin's most important contribution to ecological theory was his assertion that there were two equally important forms for the struggle for existence, namely: (1) organism against organism, which leads to competition, and (2) organism against environment, which leads to cooperation. In other words, to survive an organism must compete successfully with other organisms, but it does not compete or fight with its environment. Instead, it must adapt to and/or operate, so to speak, with its environment.

Despite the undoubted importance of mutualism (the ecological term for mutual aid), especially where resources and physical conditions are limiting, it receives very little attention in ecology textbooks. For example, Paul Keddy (1990, p. 101) tabulated the number of pages devoted to competition, predation, and mutualism in 12 textbooks published since 1975. A total of 462 pages were devoted to competition, 359 to predation, and only 59 to mutualism. It turned out that my textbook was the only one that devotes an approximately equal number of pages to each of the three interactions.

Perhaps we can get clues from territorial birds about when to confront and when to cooperate. Let us take the mockingbird as an example. The male and, to a lesser extent, its mate are very feisty and confront any other mockingbirds that attempt to invade their territory. In the bird world, territorial disputes involve a lot of singing, bluffing, scolding, displaying conspicuous feather patches (on head, wings, and tail), and chasing, but not much actual fighting. The established resident has the home field advantage and the would-be invader generally backs off if confronted vigorously.

It would seem that a certain amount of confrontation is important in maintaining order, but more important is the peace and cooperation needed to maintain the pair's home life, which involves a complex sequence of mating, nest building, egg laying, incubation, and care of the young, not to mention the time-consuming need of finding food for the family. Studies in avian energetics indicate that less than 15% of a territorial bird's time and energy is spent in confrontation. To spend more than this would result in neglect of the home life and reproduction necessary for the species to survive into the next generation.

We might suggest that the recent dramatic shift in the relationships between the superpowers is a parallel to the naturally evolved behavior of the feisty mockingbird. For several decades, the United States and the Soviet Union increased their production of weapons in the name of defense. As the expense of this confrontation reached 15 to 20% of national wealth in each country, and as the defense initiative became a powerful political means to unite people and win elections or maintain dictators, consumer, social, and environmental needs were neglected. As the pressure to address these internal needs became increasingly urgent, opportunities for moving from Cold War to cooperation were eagerly seized by both countries. As we are finding out, this transition is difficult and will take time, but it is preferable to nuclear war.

When I was teaching undergraduate ecology, I sometimes offered to give extra credit on a quiz for any student who could list the four responsibilities assigned to our federal government in the preamble of the United States Constitution. Very few students could earn this credit. Most everyone knew that "national defense" was one, and many listed "justice." But few knew that "public welfare" and "maintenance of domestic tranquility" were the other two. I believe what our founding fathers meant by the latter was maintaining a quality environment free from pollution and stress. It goes without saying that up to now we have given very little real attention and spent mighty little money on domestic tranquility as compared with what we have been doing to confront "evil empires."

To end this commentary, let us now consider the campus, ours and most others in this country. During the pioneer years, when growth was rapid, the time was right for proliferation of new buildings, departments, and schools, each competing for students. We can be proud of the fact that the University of Georgia has remained competitive with other land grant schools. But now that such rapid increases in budgets and infrastructures are less likely, we should be turning from quantity to quality growth, that is, a better but not bigger university. More cooperation between schools and departments will contribute to improving the quality of higher education and probably save money, too.

For example, the visual quality of the human-made landscape would be improved with more cooperation between architecture and environmental design (practically nonexistent on the UGA campus). The presidential task force's recommendation for an undergraduate curriculum for environmental literacy is bogged down by departmental turf battles because the administration has failed to designate a coordinator, as recommended by the task force. I could go on with many more examples, but I believe my message is clear; namely, when things get tight and there are diminishing rather than increasing returns of scale, it pays to cooperate.

The Search for the
Evil Kingdom

Eugene P. Odum

*P*eople and nations move ahead and work together when there is a common mission that motivates. Often it is a perceived common enemy that unites us. For several decades communism and the Soviet Union became our common enemy. Confronting this "evil empire" dominated our economics and politics, consumed a substantial part of our wealth, and established a military–industrial complex that remains a powerful consumptive force today. As communism has become less of a threat, the Islamic world is considered by many as the new common enemy, so we went to war with Iraq. More recently, the federal government, or government in general, seems to have become an evil force in the minds of many. Granted there are many aspects to be concerned about in all of these; focusing on a succession of common enemies in the short term tends to direct attention away from more important long-term trends and needs. I will argue that the real deadly menace, not only here but throughout the world, is *growth mania*, or more specifically, *growth beyond maintenance*, or in ecological terms, *growth beyond carrying capacity*. In other words, bigger is no longer better when the quality of human life and the life-supporting environment cannot be maintained.

When the cost of maintaining the infrastructure, the services, and dissipating the disorder that is inherent in any large complex system begins to exceed the available maintenance revenues, then *increasing* returns of scale, which

---

Unpublished essay, 1996.

economists like to talk about, will become *decreasing* returns of scale. Which is to say, too much, not too little, becomes the problem.

What is not generally understood is that the relationship between cost and size is not linear, but is a power function. For example, when a city doubles in size, the cost of maintaining services, order, and repairs more than doubles; in the real world the cost of doubling growth in size is three, or maybe four, times. This network "law" was first worked out at Bell Labs many years ago when it was discovered that the number of switches and circuits had to be more than doubled as the number of customers doubled (see Pippenger, *Sci. Am.* **238**(6):14, 1978). Stringent effort to increase efficiency can reduce maintenance costs but cannot reduce it to a simple linear function.

In terms of whole nations, geographer Karl Butzer (*Am. Scient.* **68**:17, 1980) expresses the maintenance limitation concept as follows: "Civilizations become unstable and break down when the high cost of maintenance results in a bureaucracy that makes excessive demands on the productive sector." Sound like the United States today?

Maintenance as the ultimate limitation is clearly seen in nature. Take the growth and development of a forest, for example. In a young forest (the pioneer or youthful stage, as it were), species with high reproductive potential and rapid growth potential colonize and the forest grows in size. As the timber volume grows larger, more and more of the productive energy (sun energy converted by photosynthesis) is required to maintain the leaves, trunks, roots, and animal life. At this point in development, natural selection favors species that put their energies into maintaining the individual rather than species that produce a large number of seeds or offspring. Accordingly, long-lived oaks and maples replace the short-lived rapid-growing willows, birches, or pines. When the cost of maintenance of the whole large and complex structure equals or exceeds production, then the forest stops growing in size, although the quality of the wood may continue to increase. Growth beyond maintenance in natural systems as well as in human systems becomes analogous to cancer in the individual human, that is, very deadly.

I can put this growth reality into a vignette as follows: "To grow or not to grow is not the question. The question is when to stop getting bigger and start getting better." In other words, there are times and places for quantitative growth when the choice is grow **or** die, but there are times and places when it is grow **and** die, unless that growth shifts from quantitative to qualitative. We go through this quantitative-to-qualitative sequence as individuals since we spend the first 20 years or so growing in size and complexity, and then we spend most of our life trying to become a better but not a bigger person. The transition from youth to maturity, or adolescence, is a difficult period. I suggest that our nation and much of the world is entering an analogous difficult transition.

We can put all of this in still another context. Perhaps for the first time in human history we are faced with too much of good things. Automobiles, use of natural resources, money, chemicals (fertilizers, pesticides, and drugs, for example), and people are "good things" in moderation, but can be deadly in excess. To hedge against overshoots, developed countries ought to be reducing consumption and less developed countries (and inner cities everywhere) ought to be reducing the birth rate. In the United States, we ought to be taxing consumption rather than income. If Ross Perot's suggestion of a 50¢ gasoline tax had been enacted several years ago, balancing the national budget would today not be such a traumatic effort, the air would be a bit cleaner, and fossil fuels, the lifeblood of industrial societies, would last a lot longer. Another large source of revenue that could be used for maintenance of environment and existing infrastructure is the sale of public resources such as the broadband radio spectrum for wireless communication.

The Republican's "Contract with America" is supposed to be good for our nation as a whole. So why does it contain nothing to preserve the quality of the environment, despite the fact the polls show that the majority of Americans are concerned about the environment and are willing to pay for maintaining its quality? I don't think it is because politicians are unaware of the importance of environmental protection, but rather that they, and people in general, view the environment as a separate problem, something to consider when economic and social problems have been dealt with. It's that we are accustomed to the "one problem/one solution" or the "quick-fix" approach. I would not be surprised to see a "contract for the environment" proposed in the near future when it becomes more evident that there are proposals in the Contract with America that would set back hard-earned environmental protection gains.

The reality is that the environment is not a separate problem, but is part of all human problems, whether they be public welfare, business, health, education, or whatnot. For example, environmental scarcities resulting from dividing finite space and resources among more and more people, and the concentration of wealth in the hands of a few people (i.e., the increasing gap between the rich and the poor) are certainly major causes of increased crime and violence. If we are to deal with the "evil empire" of excessive growth (and greed), we must make more holistic appraisals of our problems and predicaments.

In terms of human affairs, perhaps in contrast to nature, the maintenance limit level, or carrying capacity, is not static. It may be possible to raise the level to allow for more quantitative development by concerted action, but it cannot be done quickly by passing a law or invoking a new technology. For one thing, all technology, in the words of Paul Grey, President of MIT, "has mixed benefits," including both bright and dark sides. New technology has maintenance costs like everything else. The dark side of atomic energy technology, for example, has so far just about canceled out its prospects as a major global energy source.

Finally, we all need to remember that democracy on a large scale can never be very efficient, and that, according to Thomas Jefferson, it is the worst system we have, except for all the others. The only alternative to a democratic federal government is a military dictatorship, or as would be more likely in this country, a government dominated by the military–industrial complex. Government is not a business because there are needs that cannot be operated for profit. Environmental ethics and standards for air, water, soil, and other environmental qualities have to be nationwide, ultimately worldwide, and not just statewide. Furthermore, they now have to be raised, not lowered if we are to avoid common disasters that will come sooner or later with unmanaged growth and too much of good things.

Diversity in the Landscape: The Multilevel Approach

Eugene P. Odum

oncern about loss of "biodiversity" is reaching almost a fever pitch, if we can judge by the volume of papers and articles that are appearing not only in professional journals but in the popular press as well. Congress has even enacted legislation that gives at least token support for the preservation of biodiversity. In addition, our staid National Academy of Sciences convened one of those "quickie" conferences to assess the situation that resulted in a book (edited by E.O. Wilson, 1988) in which 38 different "experts" had their say.

Much of the concern so far has focused on the species level. But since species live in communities and require habitats and genetic fitness to survive, and since there is a great deal of natural diversity both above and below the species level, number of species (special richness) is not by itself an adequate measure. A multilevel approach is required if we are to understand and deal with diversity.

First published in *The Georgia Landscape,* The School of Environmental Design, University of Georgia, 1992.

Diversity Below The Species Level

Lifecycle stages of both plants and animals often contribute as much diversity to an ecosystem as several different species. Because they differ in feeding habits and habitat, a butterfly and its caterpillar life history stage, or a frog and its tadpole, or a tree and its seedling, provide more diversity than two closely related species of butterflies, adult frogs, or trees. One can better assess organismic diversity by counting "kinds," with life history stages counted as different "kinds," than by counting species (so one need not be a skilled taxonomist to quantify diversity!).

Genetic diversity is a special concern, and with the development of gel electrophoresis and other techniques it can now be measured in individuals and populations. As humans modify and occupy more and more space, we tend to create a "patchy" landscape, with the original habitats broken up into isolated pieces. When populations become isolated in such patches and their numbers become greatly reduced, a "genetic bottleneck" (inbreeding, genetic drift) may develop that leads to extinction. To avoid losing species diversity through such genetic bottlenecks, it is necessary to preserve large ares of habitat and/or preserve or create corridors or connections between patches so that organisms can move freely from patch to patch. For example, a string of parks established along the Georgia fall-line hills would ensure the preservation of the unique sandhill flora and fauna, many species of which are vulnerable because of narrow or restricted ranges.

Diversity Above The Species Level

Variety within trophic levels (plants, herbivores, predators, etc.) in vegetative strata and in functional niches (pollinators, for example) is important, because having more than one species capable of carrying out a necessary function in the web of life ensures that a loss of any one species won't jeopardize the whole system or landscape. We can think of this as a "redundancy diversity," or what my Georgia colleagues Hill and Wiegert call "congeneric homeostasis."

Architects and campus planners tend to favor a landscape of tall trees and grass or other ground cover with little or no shrub or understory layer (so as not to hide the beauty of the buildings, so they say!). But to leave out the shrub layer is to greatly reduce the songbird population, which to many is as an important part of the domesticated landscape as are the plants. Most of our favorite songbirds — cardinals, mockingbirds, brown thrashers, song sparrows, and so on — require shrubbery for nesting and cover. Most songbirds nest within 15 feet of the ground, not in tall trees. Accordingly, to have a high diversity of birds we need to have a diversity of vegetative strata. To remind our UGA campus planners of the desirability of shrubs, I send them from time to time a copy of a little one-page article entitled "More Birds in the Bushes from Shrubs in the Plans" that was published in *Landscape Architecture* in 1969.

Landscape Diversity

For land-use planners and landscape designers, landscape diversity is the most important aspect of diversity. We can have productive monocultures in agriculture and forestry and still maintain a high landscape diversity if the latter is accepted as a desirable goal in land-use planning and management. A diversity of natural and domesticated ecosystems is not only aesthetically pleasing but also vital to maintaining our life-support environment, which provides our increasingly urban civilization with the non-market goods and services of nature such as clean air and water, as well as food. Using the late Robert Whittaker's terminology, *alpha* or in-habitat diversity can be low, but high *beta* or between habitat diversity, and *gamma* or regional diversity, can compensate.

The landscape of the Savannah River Plant (SRP) reservation provides a good example. We recently assessed habitat diversity in this 300-square-mile area by gridding large-scale aerial photographs and recording the vegetative or other habitat type at each of 625 dots that overlaid 16 squares, totaling 4,436 acres. Each dot represented 7.1 acres. Most of the former agricultural fields and old fields, comprising about one-third of the reservation, have been converted to pine plantations. Despite the large area in monoculture, the overall habitat diversity remains high because the plantations are intermixed with numerous hedgerows, patches of natural forests, stream valleys, impoundments, and wetlands. Habitat diversity was close to natural soil diversity or about 0.7 on the evenness diversity index scale that ranges from 0 to 1. The Forest Service, which manages the timber on SRP, has proposed enlarging to pine plantations to increase pulp production, even though the cost of removing the natural vegetation and preparing areas not naturally suitable for pines is very high. We estimated that extending pine monocultures to 50% of the area would reduce diversity from 0.7 to 0.6 and that converting 75% of the land to plantations would reduce diversity to the very low level of 0.38. In this case, having no more than a third of that area in undiverse monocultures and two-thirds in a variety of natural or semi-natural ecosystems results in a very desirable landscape diversity. We suggest that this two-to-one ratio of natural to domesticated might be a model for land-use planning of large areas such as counties or states.

The Georgia Landscape

In 1987 a faculty–student task force completed a two-year study of Georgia resources and environment management as part of a comprehensive study of Georgia funded by the Kellogg Foundation. A report of this study entitled *The Georgia Landscape: A Changing Resource* is available from the Institute of Ecology. In the 50 years covered by the study (1935–85), the Georgia landscape has changed quite a bit. Cropland has decreased (twice as much food is now grown on half as many acres as in 1935), forest land has increased, and urbanization has spread. The Georgia landscape has become less "patchy" as tracts of cropland and forest land have increased in size and become more regular in shape (fractal dimensions decreased).

In a sub-study for her master's degree, Sharon Hoover compared the landscape and species diversity of vegetation in the three major physiographic regions of the state — Mountains, Piedmont, and Coastal Plain. Species diversity tended to be higher in northern Georgia, but landscape diversity proved to be higher on the Coastal Plain, due apparently to its complex drainage patterns. Her results show that diversity at one level is not necessarily the same as at another level, suggesting, of course, that to get the whole picture one must examine more than one level in a hierarchy.

At the present time, landscape diversity in Georgia is quite high; there is indeed a rich variety of habitats. Nonetheless, this favorable situation will not be sustained unless special effort is made to preserve more natural areas. Only 8% of the land area of Georgia has been set aside to remain in a natural or semi-natural condition (parks, refuges, green belts, etc.). By comparison, South Carolina has 10%, Florida 14%, and California 46%. Most of Georgia's preserved area is on the coast and in the mountains. We have been fortunate that many of the sea islands that were acquired by wealthy persons when land values were extremely low are now publicly owned. But in the Piedmont, where most Georgians live, less than 2% has been set aside as permanently preserved area.

One of the major recommendations coming out of the Kellogg study calls for a statewide effort to preserve the river corridors and freshwater wetlands. State legislation will be needed as well as promotion of conservation easements and other economic incentives as alternatives to development. Preservation of river corridors will not only enhance water quality and preserve wetlands but will also help maintain our favorable landscape diversity since stream alleys provide corridors that connect the patches.

The story of the 50-year history of the Georgia landscape has been published as Chapter 9 in the book *Changing Landscapes: An Ecological Perspective* (eds. I.S. Zonneveld and R.T.T. Forman, Springer-Verlag, New York, 1990).

Diversity and
the Survival of
the Ecosystem

Eugene P. Odum

Introduction

If we accept the general proposition or basic philosophy that life evolves as a hierarchical system, then it is convenient to consider properties of biosystems in terms of levels of organization — viz., genes, cells, organs, organisms, populations, communities, ecosystems, and so on (see, for example, Grobstein, 1969; Odum, 1975a, 1975b; Redfield, 1942). Some properties of living systems, and the methods applicable to their study, are especially relevant, or perhaps even unique, to particular levels of organization in the hierarchy. For example, birth rate is a functional property more appropriate to the population level of study than to the organism or community level. Other properties "emerge" as smaller units are integrated into larger systems, which is to say that the forest (as an ecosystem) has characteristics in addition to those of the trees (as component populations). Fiebleman (1954) has suggested that at least one major emergent property can be expected with each successive level in the hierarchical series. Still other properties are perhaps applicable over a wide range of levels. I believe diversity is a general systems property that is important at most, if not all, levels of biological

---

Originally published in *American Tissue Culture,* 50th Anniversary Symposium of the Society for Microbiology, 1976.

organization. Thus, we can consider as an important property the diversity of antibodies in an immune system as well as the diversity of species in an ecosystem.

Much has been written about genetic diversity, and especially about the consequences of reduced genetic variability in those species important to man. A massive report entitled *Genetic Vulnerability of Major Crops*, prepared by a panel of scientists and published by the National Research Council and the National Academy of Sciences (1972), concludes that the trend towards planting huge areas with the same genetic strain of food plant makes our high-energy industrialized agriculture extremely vulnerable to new epidemic diseases or climatic perturbations. The assumption is that severe oscillations in food production could result from putting all of one's eggs in the same genetic basket, as it were. As a hedge against the possible collapse of a particular monoculture, agricultural research stations are trying to maintain a variety of species and strains in nurseries or as stored seeds.

The food crop example illustrates how diversity at the species level can affect survival at a higher level of organization, the human cultural system. Diversity can be of even more direct importance at the ecosystem level in at least two major ways: (1) as a "spreading-the-risk" property that reduces the probability of extinction of a major functional group of organisms, which would then be followed by disorder in, and perhaps the death of, the system as a whole, and (2) as a major factor in the development of ecosystems from pioneer and insecure stages to more mature and stable stages (i.e., the process of ecological succession).

Spreading the Risk

A number of authors (see, for example, den Boer, 1968; Reddingius and den Boer, 1970) have discussed and set up simulation models for the stabilization of animal numbers in which a diversity of species and genetic types effectively spreads the risk of the loss of a vital community function much as insurance reduces the impact of random disasters in human affairs. As evolutionary biologists have turned from study of laboratory populations to field study, they have discovered that coexistence is more common than competitive exclusion in groups of species occupying similar niches. In other words, when the possibility of co-evolutionary adaptation of interactive species is added to the classical Lotka–Volterra competition model, then the variety of components one actually observes in nature becomes more explainable.

Community Diversity in Ecosystem Development

In addition to interactions at the species level, the processes in the ecosystems as a whole promote increased variety, not only among biological units but among physical habitats as well. The process of ecological succession can be generated in a simulation consisting of a pair of feedback loops, one positive and one negative (Gutierrez and Fey, 1975). As plant biomass accumulates in a solar-powered ecosystem on a suitable substrate, animals and microorganisms increase, thus increasing the total biomass. This, in turn, increases the diversity of niches and habitats, thereby raising the carrying capacity for more plants, animals, and microorganisms. This growth and diversification process continues until being limited by the negative feedback loop involved in regeneration of critical nutrients. In this simulation, community diversity becomes a key property in generating the developmental sequence.

Diversity as a Function of Energy Flow

In one study, the frequency distribution of diversity in a wide-ranging sample of ecosystems was found to be bimodal (Odum, 1975a, 1975b). Diversity was low in one large group of situations, in the range of 0.05 to 0.2 on a scale of 0 to 1 (where 0 represents minimum [only one kind] and 1 represents maximum diversity, where many species are of equal abundance). Another peak in the frequency curve came at moderately high diversity, 0.7 to 0.85. In no case was diversity maximum, and there were relatively few cases below the 0.5 level.

In assessing the nature of ecosystems in these two groups, it became apparent that most low-diversity systems were either subsidized with inputs of energy and/or materials, as in agricultural cultural systems, or tidal salt marshes or, alternately, they were stressed by some kind of input or perturbation. High-diversity systems were mostly those in a stable physical environment, those in a mature or advanced state of development, and resource-limited systems not receiving inputs of energy and materials from the outside. These patterns suggest that both the quantity and quality of energy flow influence the diversity level in the ecosystems as a whole. The general theory may be relevant to man's current energy crisis, as I have previously noted (Odum, 1975a):

> If we consider energy sources as very important "species," then the developed countries of the world are now in the very low diversity category with 90 to 95% dependence on fossil fuel to run cities, industries, and agriculture. Mankind has prospered in a material sense, and his population has expanded rapidly as a result of his skill in exploiting one major source of energy. Abundant high-utility fuel such as oil is a positive-feedback-forcing function that has promoted a low-diversity society whether the individual likes it or not. Now the problem is how to avoid the "bust" as this major source declines. Should mankind strive for another dominant source, such as fusion atomic power, or should he conserve, diversify, and work towards a powered-down steady state? In view of the uncertain feasibility of another dominant high-utility source, it would seem prudent for man to adopt the latter strategy.

Landscape Diversity

Since productivity and yield are enhanced by low diversity, whereas safety and perhaps stability are associated with moderately high diversity, it would be prudent for man to combine both in the future management of the earth's surface. This can be done in a land-use plan that consists of numerous blocks of a variety of systems, but with some blocks having very low internal diversity. Thus, one-third of a landscape might be devoted to monocultures of wheat or pine trees, but if blocks of these were interspersed with patches and breaks of natural habitat, then an overall diversity pattern of 75% maximum could be achieved when we consider each "block" as a "species." In an other example, a much safer, more pleasant, and easier-to-maintain situation would obtain if urban-industrial areas could be separated by stretches of undeveloped habitat along a coastline or river in contrast with a continuous "strip-city," as now too often develops in the absence of planning. The predicament of the northeastern United States reveals all too well the fact that "strip-cities" tend to overrun their life support as well as their economic resources. In conclusion, observations at several levels of organization suggest that a moderately high diversity is a desirable property of the systems of man and nature, especially when resources are fully utilized or otherwise limiting. Although diversity in natural solar-powered ecosystems seems to adjust readily and appropriately to changing conditions of energy and material input, mankind in industrial societies may need to make a special effort to promote diversity because large inputs of high-utility energy tend to drive fuel-powered ecosystems towards a low-diversity state, which undeniably increases the probability of boom-and-bust oscillations.

The Pricing
System

Eugene P. Odum

he pricing system, which is the heart of a free supply-and-demand economy, is ineffective when it comes to preserving natural environment as long as life support and other values of environment in its natural condition are not considered in making land-use decisions. Because of the irreversibility of many land-use decisions, a pricing system that considers only man-made values tends to force the development of man's fuel-powered systems beyond the optimum, that is, to a point of rapidly diminishing returns for both the developed and the necessary life-support natural environments that must be a part of man's total environment. It is clear that the time has come to extend economic accounting to include what has heretofore been considered to be the "free" work of nature, or, to put it in other words, what economists in the past have considered to be "external" values and costs (and, therefore, not included in the pricing system) must now be "internalized" to achieve a total evaluation. Whether such an extension of the pricing system will be totally effective in preserving the quality of human life and environment in a finite world remains to be seen, but it's worth a try even if only partially effective.

As a step toward realistic economic evaluation of the natural (i.e., "undeveloped") environment, we have calculated monetary values for marshlands

---

This essay is a condensed report of a monograph entitled *The Value of the Tidal Marsh* by J.G. Gosselink, E.P. Odum, and R.M. Pope, published by the Center for Wetland Resources, Louisiana State University, Baton Rouge (LSU-SG-74-03), 1974.

Basis for evaluation	Annual return per acre	Income capital-ization value per acre (at interest rate of 5%)
1. Commercial and sports fisheries	100	2,000
2. Aquaculture potential		
a. Moderate oyster culture level	630	12,600
b. Intensive oyster raft culture	1,575	31,500
3. Waste treatment		
a. Secondary	280	5,600
b. Phosphorous removal	950	19,000
c. Adjusted tertiary	2,500	50,000
4. Maximum noncompetitive Summation of values		
a. 1 + 3c	2,600	52,000
b. 2b + 3c	4,075	81,000
5. Total life-support value	4,100	82,000

Table 18.1 Marsh–estuary dollar values as determined by various methods of evaluation

and estuaries of the South Atlantic and Gulf Coasts on the basis of: (1) byproduct production (fisheries, etc.); (2) potential for aquacultural development; (3) waste assimilation; and (4) total "life-support" value in terms of the "work of nature" as a function of primary production. Money values of marsh–estuaries in their natural state were calculated in terms of (a) annual return, and (b) an income-capitalized value (equal to R/i, where R is the annual return and i is a standard interest rate of 5% (see R. Barlowe, *Land Resource Economics*, Prentice-Hall, 1965). The results of our estimates in round figures are shown in Table 18.1.

The bases for these calculations may be very briefly outlined as follows. The annual return per acre of fisheries that involves the harvest of naturally produced animals is based on summing the average dockside value of fish and shellfish, the "value added" in processing, and the Fish and Wildlife Services estimate for the value of sport fishing, and then dividing by the number of acres of estuaries that provide the nursery grounds for the harvest. Statistics from three states (Georgia, Florida, and Louisiana) were used. Estimates for oyster aquaculture is based on the potential annual yield at two levels: (1) 1800 lb oyster meat, a moderate level as actually achieved experimentally at Bear's Bluff Laboratory, in South Carolina, and (2) 4500 lbs, as achieved in intensive cultures in Japan. A non-inflationary price of 0.35 per pound of oyster meats is used in the calculations.

Values based on waste treatment work were calculated from the costs of secondary and tertiary treatment in waste treatment plants built and maintained by man. In other words, these values are what it would cost society to treat wastes in amounts not exceeding the reasonable capacity of healthy estuaries to metabolize municipal and relatively nontoxic industrial wastes, if such estuaries were not available to do this useful work. Many estuaries are actually being called upon to treat much larger amounts of domestic wastes than 3.5 pounds of BOD per day per acre, but at the expense of other uses and at a high cost in terms of deteriorated water quality. Thus, the $50,000/acre income-capitalized value (item 3c, Table 18.1) is a very conservative or minimum value representing a desirable use of natural systems for waste treatment. Mid-Atlantic estuaries such as the Potomac or Delaware would be priced far above this value in terms of their present "overtime" work to which they are being subjected. It is all too evident that we cannot continue to make nature pay all of this price without grave risk of breakdown and economic disaster to cities should they suddenly be required to double or triple taxes to pay for all this work when the "free" system breaks down. It is prudent to preserve enough nature to do a reasonable amount of waste treatment work (especially tertiary treatment, which is very expensive when done artificially), and this capacity needs to be figured into the economic value of natural environments. In making these estimates, we relied heavily on D.C. Sweet's excellent report on *The Economic and Social Importance of Estuaries* (available from the EPA, Water Quality Office, Washington, DC), and we used a cost estimate for complete tertiary treatment of $2 per pound of BOD.

Our analysis brought one important principle sharply into focus. Estuaries are not good secondary treatment systems because of the already heavy load of organic matter, which means that additional organic matter may critically affect dissolved oxygen levels, but they are superb tertiary treatment systems for nutrients. For example, very large amounts of phosphates can be discharged into an estuary where vigorous tidal action is not impeded without appreciable effect on water quality, because such an active system stores and recycles nutrients with great efficiency. Since removal of organic matter is cheap, once capital investment in the treatment plant has been made man's best strategy is to remove organic matter, together with toxic metals and synthetics, as much as possible and let nature "treat and recycle" nutrients at levels that she can handle free of charge.

Finally, the estimate of a dollar value for total life support (item 5, Table 18.1) was calculated by dividing the annual gross primary production rate expressed as kilocalories ($10,000/m^2$ or 41×104/acre, the round figure mean of most recent published estimates for Georgia and Louisiana) by 104, a round figure factor representing the number of kilocalories equivalent to one dollar (Gross Energy Consumption divided by Gross National Product). Since "productivity" is a measure of a natural system's capacity to do all kinds of useful work — such as waste treatment, CO_2 absorption, O_2 production, raising shrimp, supporting waterfowl, protecting cities and beaches from storms, providing transportation, and so on — then converting work energy to money is a quick and convenient way of evaluating a given natural system. This approach has the great advantage that it can be applied to a specific acre (thus high marshes and low marshes can be valued differently according to their respective capacities to do useful work), while the other approaches discussed in this paper apply only to large estuarine systems as a whole.

The value of estuaries for waste assimilation and general life support is far greater than that accruing from byproducts, as would be expected, since the latter represents only a fraction of work energy potential. In the past, conservationists have attempted to justify preservation on an economic value basis of such byproduct uses, often inflating such values beyond reality. It is obvious that such partial evaluation not only cannot match real estate values, but they represent only a fraction of the real value of natural environment. Extremely intensive aquaculture, such as raft culture of oysters, would make estuaries economically more attractive, but it is not a good option for Georgia because of the large energy and labor subsidies required. Also, if aquaculture were carried out on a large scale, it would conflict with many other desirable functions. Aquaculture certainly has its place in the future but should not be considered to be the sole economic reason for preserving estuaries.

Summing values for components that could conceivably be noncompetitive gives a "multiple-use" value approaching that based on productivity (Table 18.1, compare items 4 and 5). In a sense, this validates the theoretical principle that the natural productivity rate is an index of the useful work potential of a natural ecosystem.

There are, of course, still other values of an aesthetic and scenic nature that are not included in these economic evaluations. While most of us will agree that aesthetic values can never be adequately translated into dollars, there are approximations that can be helpful. For example, I recently asked a developer who is trying to "design with nature," as far as possible, to estimate the price he will charge for property with a view of the marshes and waterways, as compared with property without such a view. He replied that the former would be priced about twice that of the latter. Thus, we could use this increment in calculating "value added" for a permanently preserved natural area that is adjacent to developed residential property.

Demonstrating that marshlands and estuaries have a substantial dollar value in their natural state certainly provides a big boost to preservation of such areas that are in public ownership. If large values such as those in Table 18.1 (items 4 and 5) are generally recognized and accepted, then state or federal agencies, or commissions which have jurisdiction over the property or resource, will be less likely to lease, give away, or sell valuable marshlands for capricious development. Also, planners will have greater incentive and public support for zoning such areas into permanent protective categories.

On the other hand, if the marshland is in private ownership, the owner will stand to gain by selling for development no matter how high the appraisal, since leaving the area in its natural state earns the owner little or no return. Floodplains and river swamp pose a similar dilemma throughout the state. The dichotomy of interests between the value to the owner and the value to society becomes an increasingly serious problem as population growth and industrial development accelerate. The pricing system, as it now operates, as we have already indicated, offers no solution to this problem since development becomes essentially an irreversible action. Thus, even though the value of marshland increases as it becomes scarcer to an eventual point that its life-support value could "outbid" other land uses, there is no way to convert the previous development back to its former (and now more valuable) state. The irony of dependence on the price system is that it can make a reasonable-sounding argument for developing marshland, and it can even offer an argument that a point will be reached when the land should be converted back to marsh, but it cannot effectively recreate marshland, a very expensive process, even if technically possible.

It is worth mentioning that, as high as the values herein determined for marsh acreage may seem to be, these values are based on availability of large areas of marsh and a moderate level of human population density and industrial development. All values will tend *to increase with each increment of marsh lost to an alternative use, as well as through increases in population and industrial development.* Less acreage in natural marsh doing the same or more work for man than is now done would indicate a higher value per acre to society, but it is apparent that sooner or later a limit would be reached beyond which further reduction of marsh–estuary acreage may prove disastrous. This point has certainly been reached in the Middle Atlantic states. The people of Georgia are to be congratulated in their foresight in establishing a strong marshland protection act before real estate speculation had driven the price up to a point of forcing development, even where not desired by the citizens who must pay the costs of inappropriate uses. This state action is an investment that must be guarded by vigilant public opinion like so much gold in the bank.

Evaluation of marshland as a renewable resource, e.g., as an income stream stretching into the future and increasing continually, represents one way to alleviate the destructive tendency inherent in the pricing system as it now operates. The time has come to seek ways to let the owners of natural resources with value to society receive a return. Direct purchase by government is one solution, of course; scenic or open-space easement and tax relief are other approaches. Setting up wetland "banks" where the owner is paid (or relieved of tax burden) not to develop (as in "soil banks") is perhaps a feasible "delayed option" procedure in cases where outright purchase cannot be made at a particular time.

The best solution is a "look ahead" land-use plan that delimits the amount and location of life-support natural areas that will be necessary to support a future desirable level of population density, industrial, and recreational development. Such areas can then be acquired or zoned into the public domain before the spiral of land speculation raises the market price beyond reason. We can begin to see that the present stop-gap method of making impact assessments for each and every proposed development (as now required by the NEPA Act) must evolve as rapidly as possible into regional land-use planning.

Energy, Ecology, and Economics

Howard T. Odum

Many are beginning to see that energy, ecology, and economics form a single, unified system, states the author, who gives twenty points to explain the energy control of our economy and the relationship to the environment. The net reserves of fossil fuels are mainly unknown, he says, but they are much smaller than the gross reserves that have been the basis of public discussions and decisions that imply that growth can continue. He offers a general answer to the present world situation, where "boom-and-bust" economies may soon be forced toward a steady state: reject economic expansionism, stop growth, use available energies for cultural conversion to steady state, and seek out the condition now that will come anyway.

A s long-predicted energy shortages appear, as questions about the interaction of energy and environment are raised in legislatures and parliaments, and as energy-related inflation dominates public concern, many are beginning to see that there is a unity of the single system of energy, ecology, and economics. The world's leadership, however, is mainly advised by specialists who study only one part of the system at a time.

Instead of a single system's understanding, we have adversary arguments dangerous to the welfare of nations and the role of man as the earth's information bearer and programmatic custodian. Many economic models

Reprinted with permission from *AMBIO*, published by the Royal Swedish Academy of Sciences, Stockholm, 1972.

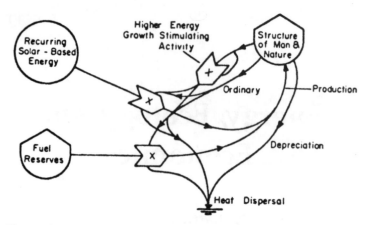

Figure 19.1A   Generalized world model of man and nature based on one-shot fossil fuel usages and steady solar work. Pathways are flows of energy from outside source (circle) through interactions (pointed blocks marked "X" to show multiplier action) to final dispersion of dispersed heat. The tank symbol refers to storage. Here world fuel reserve storage helps build a storage of structure of man's buildings, information, population, and culture.

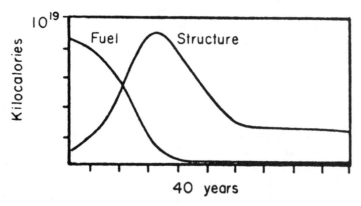

Figure 19.1B   Graphs resulting from simulation of the model in Figure 1A. Available world fuel reserve was taken as 5×10^{19} kilocalories and energy converted from solar input, and converged into man's productive system of growth and maintenance was 5×10^{16} kilocalories when structure was 10^{18} kilocalories. Peak of structural growth was variable over a 50-year period depending on amounts diverted into waste pathways.

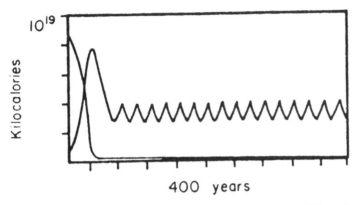

400 years

Figure 19.1C   The steady state observed in some simulations of Figure 1A
was an oscillating one, as in the graph shown here.

ignore the changing force of energy regarding the effects of energy sources
as an external constant; ecoactivists cause governments to waste energy in
unnecessary technology; and the false gods of growth and medical ethics
make famine, disease, and catalytic collapse more and more likely for much
of the world. Some energy specialists consider the environment as an antago-
nist instead of a major energy ally in supporting the biosphere.

Instead of the confusion that comes from western civilization's characteristic educational approach of isolating variables in tunnel-vision thinking,
let us here seek a common sense overview that comes from overall energetics. Very simple overall energy diagrams clarify issues quantitatively, indicating what is possible. The diagrams and symbols are explained further in a
recent book (Odum, 1971).

For example, Figure 19.1 shows the basis of production in interaction of
fuel reserves, steady energies of solar origin, and feedback of work from the
system's structure. Figure 19.1 is a computer simulation of this model for our
existence, showing a steady state after our current growing period. As the fuel
tank is drained, we return to a lower solar base of simpler agriculture. Simple
macroscopic minimodels based on an overview of world energy provides the
same kind of trend curves as the detailed models of Forrester and Meadows
(see Meadows et al., 1972). With major changes confronting us, let us
consider here some of the main points that we must comprehend so we may
be prepared for the future.

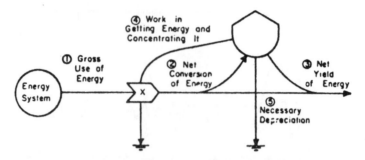

Figure 19.2   Energy flow diagram illustrating energy laws, and the difference between net and gross energy flows.

1. The true value of energy to society is the net energy, which is that after the energy costs of getting and concentrating that energy are subtracted

Many forms of energy are low grade because they have to be concentrated, transported, dug from deep in the earth, or pumped from far at sea. Much energy has to be used directly and indirectly to support the machinery, people, supply systems, etc., to deliver the energy. If it takes ten units of energy to bring ten units of energy to the point of use, then there is no net energy. Right now we dig further and further, deeper and deeper, and go for energies that are more and more dilute in the rocks. Sunlight is also a dilute energy that requires work to harness.

We are still expanding our rate of consumption of gross energy, but since we are feeding a higher and higher percentage back into the energy seeking process, we are decreasing our percentage of net energy production. Many of our proposed alternative energy sources take more energy feedback than present processes. Figure 19.2 shows net energy emerging beyond the work and structural maintenance costs of energy processing.

2. Worldwide inflation is driven in part by the increasing fraction of our fossil fuels that have to be used in getting more fossil fuel and other fuels

If the money circulating is the same or increasing, and if the quality of energy reaching society for its general work is less because so much energy has to go immediately into the energy-getting process, then the real work to society

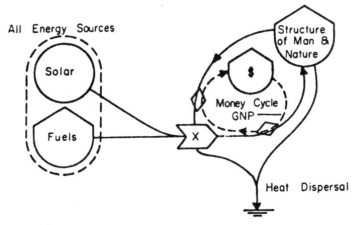

Figure 19.3   Relationship of money cycles to the energy circuit loops.

per unit money circulated is less. Money buys less real work of other types, and thus money is worth less. Because the economy and total energy utilization are still expanding, we are misled to think that total value is expanding, and we allow more money to circulate that makes the money-to-work ratio even larger. Figure 19.3 shows the circulation of money that constitutes the GNP in a countercurrent to the energy flow.

3. Many calculations of energy reserves that are supposed to offer years of supply are as gross energy rather than net energy and thus may be of much shorter duration than often stated

Suppose for every ten units of some quality of oil shale proposed as an energy source there were required nine units of energy to mine, process, concentrate, transport, and meet environmental requirements. Such a reserve would deliver 1/10 as much net energy and last 1/10 as long as was calculated. Leaders should demand of our estimators of energy reserves that they make their energy calculations in units of net energy. The net reserves of fossil fuels are mainly unknown, but they are much smaller than the gross reserves, which have been the basis of public discussions and decisions that imply that growth can continue.

4. Societies compete for economic survival by Lotka's principle, which says that systems win and dominate that maximize their useful total power from all sources and flexibly distribute this power toward needs affecting survival

The programs of forests, seas, cities, and countries survive that maximize their system's power for useful purposes. The first requirement is that opportunities to gain inflowing power be maximized, and the second requirement is that energy utilization be effective and not wasteful as compared to competitors or alternatives. For further discussion, see Lotka (1922) and Odum (1971).

5. During times when there are opportunities to expand one's power inflows, the survival premium by Lotka's principle is on rapid growth even though there may be waste

We observe dog-eat-dog growth competition every time a new vegetation colonizes a bare field where the immediate survival premium is first placed on rapid expansion to cover the available energy receiving surfaces. The early-growth ecosystems put out weeds of poor structure and quality, which are wasteful in their energy-capturing efficiencies but effective in getting growth even though the structures are not long-lasting. Most recently, modern communities of man have experienced 200 years of colonizing growth, expanding to new energy sources such as fossil fuels, new agricultural lands, and other special energy sources. Western culture and, more recently, Eastern and Third World cultures are locked into a mode of belief in growth as necessary to survival. "Grow or perish" is what Lotka's principle requires, but only during periods when there are energy sources that are not yet tapped. Figure 19.3 shows the structure that must be built in order to be competitive in processing energy.

6. During times when energy flows have been tapped and there are no new sources, Lotka's principle requires that those systems win that do not attempt fruitless growth but instead use all available energies in long-staying, high-diversity, steady-state works

Whenever an ecosystem reaches its steady state after periods of succession, the rapid net growth specialists are replaced by a new team of higher diversity, higher quality, and longer living. Collectively, through division of labor and specialization, the climax team gets more energy out of the steady flow of available source energy than those specialized in fast growth could.

Our system of man and nature will soon be shifting from rapid growth as the criterion of economic survival to steady-state non-growth as the criterion of maximizing one's work for economic survival (Figure 19.1). The timing depends only on the reality of one or two possibly high-yielding nuclear energy processes (fusion and breeder reactions), which may or may not be very yielding.

Ecologists are familiar with both growth states and steady state, and observe both in natural systems in their work routinely, but economists were all trained in their subject during rapid growth and most do not even know there is such a thing as steady state. Most economic advisors have never seen a steady state, even though most of man's million-year history was close to steady state. Only the last two centuries have seen a burst of temporary growth because of temporary use of special energy supplies that accumulated over long periods of geologic time.

7. High quality of life for humans and equitable economic distribution are more closely approximated in steady state than in growth periods

During growth, emphasis is on competition, and large differences in economic and energetic welfare develop; competitive exclusion, instability, poverty, and unequal wealth are characteristic. During steady state, competition is controlled and eliminated, being replaced with regulatory systems, high division and diversity of labor, uniform energy distributions, little change, and growth only for replacement purposes. Love of stable system quality replaces love of net gain. Religious ethics adopt something closer to that of those primitive peoples that were formerly dominant in zones of the world with cultures based on the steady energy flows from the sun. Socialistic ideals about distribution are more consistent with steady state than growth.

8. The successfully competing economy must use its net output of richer quality energy flows to subsidize the poorer-quality energy flow so that total power is maximized

In ecosystems, diversity of species develop that allow more of the energies to be tapped. Many of the species that are specialists in getting lesser and residual energies receive subsidies from the richer components. For example,

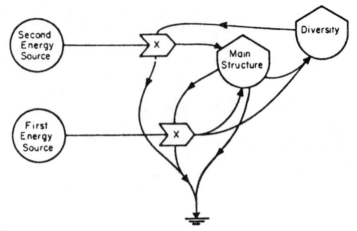

Figure 19.4 Relationship of general structural maintenance to diversity and secondary energy sources.

the sun leaves transport fuels on the tops of trees that help the shaded leaves so they can get some additional energy from the last rays of dim light reaching the forest floor. The system that uses its excess energies in getting a little more energy, even from sources that would not be net yielding alone, develops more total work and more resources for total survival. In similar ways, we now use our rich fossil fuels to keep all kinds of goods and services of our economy cheap so that the marginal kinds of energies may receive the subsidy benefit that makes them yielders, whereas they would not be able to generate much without the subsidy. Figure 19.4 shows the role of diversity in tapping auxiliary energies and maintaining flexibility to changing sources.

9. Energy sources that are now marginal, being supported by hidden subsidies based on fossil fuel, become less economic when the hidden subsidy is removed

A corollary of the previous principle of using rich energies to subsidize marginal ones is that the marginal energy sources will not be as net yielding later, since there will be no subsidy. This truth is often stated backwards in economists' concepts because there is inadequate recognition of external changes in energy quality. Often they propose that marginal energy sources will be economic later when the rich sources are gone. An energy source is not a source unless it is contributing yields, and the ability of marginal sources to yield goes down as the other sources of subsidy become poorer. Figure 19.4 shows these relationships.

10. Increasing energy efficiency with new technology is not an energy solution, since most technological innovations are really diversions of cheap energy into hidden subsidies in the form of fancy, energy-expensive structures

Most of our century of progress with increasing efficiencies of engines has really been spent developing mechanisms to subsidize a process with a second energy source. Many calculations of efficiency omit these energy inputs. We build better engines by putting more energy into the complex factories for manufacturing the equipment. The percentage of energy yield in terms of all the energies incoming may be less, not greater. Making energy net yielding is the only process not amenable to high energy-based technology.

11. Even in urban areas more than half of the useful work on which our society is based comes from the natural flows of sun, wind, waters, waves, etc., that act through the broad areas of seas and landscapes without money payment; an economy, to compete and survive, must maximize its use of these energies, not destroying their enormous free subsidies; the necessity of environmental inputs is often not realized until they are displaced

When an area first grows, it may add some new energy sources in fuels and electric power, but when it gets to about 50% of the area developed it begins to destroy and diminish as much necessary life-support work that was free and unnoticed as it adds. At this point, further growth may produce a poor ability in economic competition because the area now has higher energy drains. For example, areas that grow too dense with urban developments may pave over the areas that formerly accepted and reprocessed wastewaters. As a consequence, special tertiary waste treatments become necessary and monetary, and energy drains are diverted from useful works to works that were formerly supplied free.

12. Environmental technology that duplicates the work available from the ecological sector is an economic handicap

As growth of urban areas has become concentrated, much of our energies and research-and-development work has been going into developing energy-costing technology to protect the environment from wastes, whereas most wastes are themselves rich energy sources for which there are, in most cases, ecosystems capable of using and recycling wastes as a partner of the city without a drain on the scarce fossil fuels. Soils take up carbon monoxide, forests absorb nutrients, swamps accept and regulate floodwaters. If growth is so dense that environmental technology is required, then it is too dense to be economically vital for the combined system of man and nature there. The growth needs to be arrested, or it will arrest itself with the depressed poorly competing economy of man and of his environs. For example, there is rarely an excuse for tertiary treatment because there is no excuse for such dense packing of growth that the natural buffer lands cannot be a good cheap recycling partner. Man as a partner of nature must use nature well, and this does not mean to crowd it out and pave it over; nor does it mean developing industries that compete with nature for the waters and wastes that would be an energy contributor to the survival of both.

13. Solar energy is very dilute, and the inherent energy cost of concentrating solar energy into a form for human use has already been maximized by forests and food producing plants; without energy subsidy there is no yield from the sun possible beyond the familiar yields from forestry and agriculture

Advocates of major new energies available from the sun don't understand that the concentration quality of solar energy is very low, being only 10^{-16} kilocalories per cubic centimeter. Much of this has to be used up in upgrading to food quality. Plants build tiny microscopic semiconductor photon receptors that are the same in principle as the solar cells advocated at vastly greater expense by some solar advocates. The plants have already maximized use of sunlight, by which they support an ecosystem whose diverse work helps maximize this conversion, as shown in Figure 19.5A. If man and his work are substituted for much of the ecosystem so that he and his farm animals do the recycling and management, higher yield results, as in sacred cow agriculture (Figure 19.5B). Higher yields require large fossil fuel subsidies in doing some of the work. For example, making the solar receiving structures (Figure 19.5C), whereas the plants and ecosystem make their equipment out of the

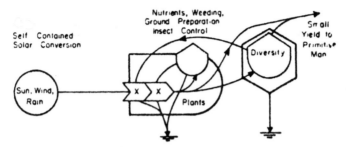

Figure 19.5A   Man a minor part of the complex forest ecosystem.

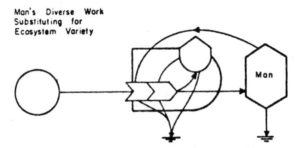

Figure 19.5B   Man a major partner in an agricultural system on light alone.

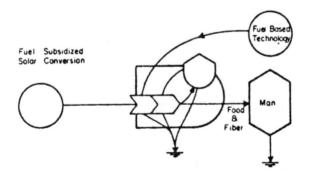

Figure 19.5C   Fossil fuel-subsidized agriculture as a colonial member of a technological society of man with maximum possible solar conversion.

Figure 19.5   Diagrams of three systems of solar energy use.

energy budget they process. Since man has already learned how to subsidize agriculture and forestry with fossil fuels when he has them, solar technology becomes a duplication. The reason major solar technology has not and will not be a major contributor to a substitute for fossil fuels is that it will not compete without energy subsidy from the fossil-fuel economy. Some energy savings are possible in house heating on a minor scale.

14. Energy is measured by calories, BTUs, kilowatt hours, and other intraconvertible units, but energy has a scale of quality that is not indicated by these measures

The ability to do work for man depends on the energy quality and quantity, and this is measurable by the amount of energy of a lower quality grade required to develop the higher grade. The scale of energy goes from dilute sunlight up to plant matter to coal, from coal to oil to electricity, and up to the high-quality efforts of computer and human information processing.

15. Nuclear energy is now mainly subsidized with fossil fuels and barely yields net energy

High costs of mining, processing fuels, developing costly plants, storing wastes, operating complex safety systems, and operating government agencies make present nuclear energy one of the marginal sources that add some energy now, while they are subsidized by a rich economy. A self-contained isolated nuclear energy does not now exist. Since the present nuclear energy is marginal while it uses the cream of rich fuels accumulated during times of rich fossil fuel excess, and because the present rich reserves of nuclear fuel will last no longer than fossil fuels, there may not be a major long-range effect of present nuclear technology on economic survival. The high energy cost of nuclear construction may be a factor accelerating exhaustion of the richer fuels. Figure 19.4 illustrates this principle.

The Breeder Process: The breeder process is now being given its first tests for economic effectiveness, and we don't yet know how net yielding it will be. The present nuclear plants are using up the rich fuels that could support the breeder reactors if these turn out to be net yielders over and beyond the expected high energy costs in safety costs, occasional accidents, reprocessing plants, etc. Should we use the last of our rich fossil fuel wealth for the high research-and-development costs and high capital investments of processes too late to develop a net yield?

Fusion: The big question is this: Will fusion be a major net yield? The feasibility of pilot plants with the fusion process is unknown. There is no knowledge yet as to the net energy in fusion or the amounts of energy subsidy that fusion may require. Because of this uncertainty, we cannot be sure about the otherwise sure leveling and decline in total energy flows that may soon be the pattern for our world.

16. Substantial energy storages are required for stability of an economy against fluctuations of economies, or of natural causes, and of military threats

The frantic rush to use the last of the rich oil and gas that is easy to harvest for a little more growth and tourism is not the way to maintain power stability or political and military security for the world community of nations as a whole. World stability requires a de-energizing of the capabilities for vast war, and an evenly distributed power base for regular defense establishments that need to be evenly balanced without great power gradients that encourage changes in military boundaries. Two-year storage is required for stability of a component.

17. The total tendency for net favorable balance of payments of a country relative to others depends on the relative net energy of that country, including its natural and fuel-based energies minus its wastes and nonproductive energy uses

Countries with their own rich energies can export goods and services with less requirement for money than those that have to use their money to buy their fuels. Those countries with interior energy flows into useful work become subordinate energy dependent on other countries. A country that sells oil but does not use it within its boundaries to develop useful work is equally subordinate since a major flow of necessary high-quality energy in the form of technical goods and services is external in this case. The country with the strongest position is the one with a combination of internal sources of rich

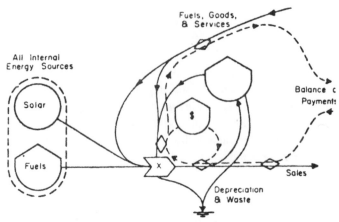

Figure 19.6   Diagram showing how energy sources and energy loss pathways affect the balance of payments and general economic competition position of a single country. Better balance results when one's own energy sources are better, and one's waste less.

energies and internal sources of developed structure and information based on the energy. The relationships of energy sources to payment balances are depicted in Figure 19.6.

18. During periods of expanding energy availabilities, many kinds of growth-priming activities may favor economic vitality and the economy's ability to compete; institutions, customs, and economic policies aid by accelerating energy consumption in an autocatalytic way

Many pump-priming properties of fast-growing economies have been naturally selected and remain in procedures of government and culture. Urban concentrations, the high use of cars, economic subsidy to growth, oil depletion allowances, subsidies to population growth, advertising, high-rise building, etc., are costly in energy for their operation and maintenance, but favor economic vitality as long as their role as pump primers is successful in increasing the flow of energy over and beyond their special cost. Intensely concentrated densities of power use have been economic in the past because their activities have accelerated the system's growth during a period when there were new energy sources in the environmental compass.

19. During periods when expansion of energy sources is not possible, then the many high-density and growth-promoting policies and structures become an energy liability because their high energy cost is no longer accelerating energy yield

The pattern of urban concentration and the policies of economic growth simulation that were necessary and successful in energy growth competition periods are soon to shift. There will be a premium against the use of pump-priming characteristics since there will be no more unpumped energy to prime. What worked before will no longer work, and the opposite becomes the pattern that is economically successful. All this makes sense and is commonplace to those who study various kinds of ecosystems, but the economic advisors will be sorely pressed and lose some confidence until they learn about the steady state and its criteria for economic success. Countries with great costly investments in concentrated economic activity, excessive transportation customs, and subsidies to industrial expansion will have severe stresses. Even now the countries that have not gone so far in rapid successional growth are setting out to do so at the very time when their former more steady-state culture is about to begin to become a more favored economic state comparatively.

20. Systems in nature are known that shift from fast growth to steady state gradually with programmatic substitution, but other instances are known in which the shift is marked by total crash and destruction of the growth system before the emergence of the succeeding steady-state regime

Because energies and monies for research, development, and thinking are abundant only during growth and not during energy leveling or decline, there is a great danger that the means for developing the steady state will not be ready when they are needed, which may be no more than 5 years away but is probably more like 20 years. (If fusion energy is a large net energy yielder, there may be a later growth period when the intensity of human power development begins to affect and reduce the main life-support systems of the oceans, atmospheres, and general biosphere.)

The humanitarian customs of the earth's countries now in regard to medical aid, famine, and epidemic are such that no country is allowed to develop major food and other critical energy shortages because the others rush in with their reserves. This practice had ensured that no country will starve in a major way until we ail starve together when the reserves are no longer there.

Chronic disease has evolved with man as its regulator, being normally a device for infant mortality and merciful old age death. It provided on the average an impersonal and accurate energy testing of body vitalities, adjusting the survival rate to the energy resources. Even in the modern period of high-energy medical miracles, the energy for total medical care systems is a function of the country's total energies, and as energies per capita fall again so will the energy for medicine per capita, and the role of disease will again develop its larger role in the population regulation system. Chronic disease at its best was and is a very energy-inexpensive regulator.

Epidemic disease is something else. Nature's systems normally use the principle of diversity to eliminate epidemics. Vice versa, epidemic disease is nature's device to eliminate monoculture, which may be inherently unstable. Man is presently allowed the special high yields of various monocultures, including his own high-density population, his paper source in pine trees, and his miracle rice, only so long as he has special energies to protect these artificial ways and substitute them for disease that would restore the high-diversity system, ultimately the more stable flow of energy.

The terrible possibility that is before us is that there will be continued insistence on growth with our last energies by the economic advisors that don't understand, so that there are no reserves with which to make a change, to hold order, and to cushion a period when populations must drop. Disease-borne reduction of man and of his plant production systems could be planetary and sudden if the ratio of population to food and medical systems is pushed to the maximum at a time of falling net energy. At some point, the great gaunt towers of nuclear energy installations, oil drilling, and urban cluster will stand empty for lack of enough fuel technology to keep them running. A new class of dinosaurs will have passed away. Man will survive as he reprograms readily to that which the ecosystem needs of him, so long as he does not forget who is serving whom. What is done well for the ecosystem is good for man. However, the cultures that say only what is good for man is good for nature may pass away and be forgotten like the rest.

There was a famous theory in paleoecology called orthogenesis, which suggested that some of the great animals of the past were part of systems that were locked into evolutionary mechanisms by which the larger ones took over from smaller ones. The mechanisms then became so fixed that they carried the size trend beyond the point of survival, whereupon the species went extinct. Perhaps this is the main question of ecology, economics, and energy. Has the human system frozen its direction into an orthogenetic path toward cultural crash, or is the great creative activity of the current energy-rich world already sensing the need for change? Are alternatives already being tested by our youth so they will be ready for the gradual transition to a fine steady state that carries the best of our recent cultural evolution into new, more miniaturized, more dilute, and more delicate ways of man–nature?

In looking ahead, the United States and some other countries may be lucky to be forced by changing energy availabilities to examine themselves, level their growth, and change their culture towards the steady state early enough so as to be ready with some tested designs before the world as a whole is forced to this. A most fearful sight is the behavior of Germany and Japan, who have little alternative energies and rush crazily into a boom-and-bust economy on temporary and borrowed pipelines and tankers, throwing out what was stable and safe to become rich for a short period; monkey see, monkey do. Consider also Sweden, which once before boomed and busted in its age of Baltic ships while cutting its virgin timber. Later it was completely stable on water power and agriculture, but now, after a few years of growth, has become like the rest, another bunch of engines on another set of oil flows, culture that may not be long for this world.

What is the general answer? Eject economic expansionism, stop growth, use available energies for cultural conversion to steady state, seek out the condition now that will come anyway, but by our service be our biosphere's handmaiden anew.

Prerequisites for Sustainability

Eugene P. Odum

*J*f we are thinking about sustainability in terms of maintaining the quality of life for humans and the environment, then it is self-evident that there have to be major changes in the way we think, behave, and do business. "Quick fixes" such as energy taxes, paper recycling, or reduction of ozone-destroying chemicals that are now being undertaken will only postpone the day when more fundamental changes will have to be evolved. In this commentary, I present four of what I call "conceptual models," that is, statements or diagrams that depict a unifying concept or paradigm that I believe is especially relevant to sustainability.

ONE: Non-market (life-support) goods and services must be incorporated into the economic system

Current free-market economics deals almost exclusively with human-made goods and services, while nature's goods and services that provide our physiological needs for breathing, drinking, and eating — along with aesthetics — are largely unpriced and not appreciated until there are serious declines in the quality of these services. At the present time, we rely on political and legal means to protect our life-support resources, but such efforts are increasingly too little and too late. Somehow, future capitalism

---

Reprinted from *Biosphere Newsletter* 3(3):10–11, 1993.

must be based on the integration of nature's capital with human capital, since each is ultimately dependent on the other.

It is encouraging that both economists and ecologists are seeking ways to incorporate non-market values into the economic system. During the past five years, there has been an outpouring of books, journals, and articles dealing with the ecology–economics interface. One thing is certain: unless life-support resources — i.e., atmosphere, oceans, wetlands, soils, forests, croplands, and so on — are properly valued and protected, sustainable development cannot be achieved, and the quality of human life will decline as human populations and their demands increase.

TWO: Sigmoid (constrained, self-regulating, or managed) growth must replace exponential (boom-and-bust) growth

Overshooting basic limits, e.g., resources, space, pollution, maintenance costs, and so on, is almost inevitable if growth continues exponentially, e.g., doubling at short time intervals. On the other hand, with sigmoid growth (the S-curve) the growth rate is progressively slowed as limits are approached. Only when growth is managed in this way can sustainability be achieved.

First and foremost, human population growth must be brought under sigmoid control, especially where rates are 2% or more, i.e., doubling every 35 years or less. Affluence and education is reducing the population growth rate in some rich countries, but in the poorer countries, which comprise two-thirds of the world population, this "demographic transition" is not working because of high levels of poverty and low levels of education. Worldwide, more direct measures involving family planning and wide use of new birth control technology are necessary if there is any hope of population leveling off early in the next century.

Second and equally important, quantitative economic growth, which involves getting larger, must be replaced by qualitative economic development, which involves getting better, as saturation levels of life support and resource use are approached. As suggested in a 1991 UNESCO report, this transition should come first in developed countries where per-capita deleterious impact on the environment is very high, and later in developing countries, where traditional economic growth may still be needed to reduce poverty. As far as anyone knows, there are no limits to quality development, but there are certainly limits to economic growth that are based on increasing consumption in a finite world.

THREE: *Land-use planning must become a major consideration*
in development

In order to sustain our life-support systems and avoid the dreaded "dooms-day" boom and bust, serious land-use planning must become a major priority at the regional as well as local level. Natural areas and agricultural lands adequate to maintain the quality of urban-industrial development must be set aside by purchase, easements, or zoning restrictions before real estate value becomes prohibitive.

What is not known at the present time is how much "set-asides" are needed: two hectares (five acres) per capita or 20–30% of the landscape have been suggested. Much depends on lifestyle. Here is where the Biosphere II experiments can help. A major mission of the first experiment completed in September 1993 was to find out if one hectare of pollution-free landscape (that is, no automobiles or industries) consisting of 80% natural and semi-natural ecosystems (the biomes) and 20% labor-intensive polyculture agri-culture would provide bioregenerative support for eight people living at a subsistence level. The answer apparently is "just barely." It becomes evident that with the lifestyle of an American with two or more cars per family and an annual per-capita energy consumption of 80 million kilocalories (com-pared to 2 million for a Pakistani), a great deal more life-support environment per capita would be needed.

FOUR: *When resources are scarce in terms of need or demand,*
then it pays to cooperate

One thing we learn from the study of whole ecosystems is that, when nutrients or other vital resources become scarce, as for example in pacific coral reefs, tropical rainforests on sandy soil, or pine forests on eroded land, a great deal of obligate mutualism evolves. For example, partnership be-tween producers and consumers (i.e., entozoic algae and the animal polyp in a coral, or roots and mycorrhizal fungi in a forest) enable these systems not only to survive but to prosper in a world of scarce resources. We humans need to learn how to do likewise, since many resources will be in short supply no matter what we do to slow down growth and demands in the future. The shift from confrontation to cooperation among the superpowers is a parallel to the natural development of mutualism when things get tough. Also, it is encour-aging that business is showing signs of moving from cutthroat competition to cooperation on the global scale.

The Biosphere II experiment reaffirms the survival value of cooperation when resources are scarce. The eight biospherians were not only able to survive for two years in the restricted enclosure but to come out in good health and still speaking to each other because they were selected and trained, not as scientists, but as resourceful individuals who could work together for mutual benefit.

Summary

No one expects the major changes we have been discussing to come quickly, but come they must, or else humanity will be doomed to trying to survive in an overcrowded and poisoned world. Transitions will take time, and in many cases will be painful in terms of restrictions and tax burdens that may have to be imposed. We have some time to make changes since the earth is resilient, and as a whole, is not yet badly damaged. We have, or can develop, technology to repair damaged components if we first reduce the stress on them. I am optimistic that we can make the transition from youth to maturity, as it were, because of: (1) the increased attention environmental matters are receiving in the media and in education, and (2) the small starts, as noted in this commentary, that are being made to bring humans and the biosphere into better harmony. I, for one, believe that the quality of life for humans and nature will be much improved in a more relaxed, less competitive, more mutualistic, and less consumptive future world.

Quantitative economic growth, which involves getting larger, must be replaced by qualitative economic development, which involves getting better.

Environmental Degradation and the Tyranny of Small Decisions

William E. Odum

*E*conomist Alfred E. Kahn's premise of "the tyranny of small decisions" is applicable to environmental issues. Examples of so-called "small decision effects" range from loss of prime farmland and acid precipitation to mismanagement of the Florida Everglades. A holistic rather than reductionist perspective is needed to avoid the undesirable cumulative effects of small decisions (Kahn, 1966).

Ideally, society's problems are resolved through a system of nested levels of public decisions. At the lowest level, decisions are made by the individual or by small groups of individuals. Higher decision-making levels range from local and state governments to the highest levels of the federal government. Theoretically, the highest levels are composed of experts whose joint decisions provide constraints in the form of "rules" for decisions made at the lower levels. Unfortunately, important decisions are often reached in an entirely different manner. A series of small, apparently independent decisions are made, often by individuals or small groups of individuals. The end result

Reprinted with permission from *Bioscience* **32**:728–729, © 1982 American Institute of Biological Sciences.

is that a big decision occurs (post hoc) as an accretion of these small decisions; the central question is never addressed directly at the higher decision-making levels. Usually, this process does not produce an optimal, desired, or preferred solution for society.

This process of post hoc decision-making has been termed "the tyranny of small decision" by Kahn (1966). As he has pointed out, this is a common problem in market economics. He gives as an example the loss of passenger train service to Ithaca, New York. Even though the majority of the inhabitants of Ithaca would have preferred to retain passenger train service, they "decided" to terminate service through the combined effects of a series of small independent decisions to travel by automobile, airplane, and bus.

Small Decisions and the Environment

Clearly, "the tyranny of small decisions," or what might be called "small decision effects," applies to much more than market economics. Much of the current confusion and distress surrounding environmental issues can be traced to decisions that were never consciously made, but simply resulted from a series of small decisions. Consider, for example, the loss of coastal wetlands on the East Coast of the United States between 1950 and 1970. No one purposely planned to destroy almost 50% of the existing marshland along the coasts of Connecticut and Massachusetts. In fact, if the public had been asked whether coastal wetlands should be preserved or converted to some other use, preservation would probably have been supported. However, through hundreds of little decisions and the conversion of hundreds of small tracts of marshland, a major decision in favor of extensive wetlands conversion was made without ever addressing the issue directly.

Regional problems are highly vulnerable to small decision effects. The ecological integrity of the Florida Everglades has suffered, not from a single adverse decision, but from a multitude of small pinpricks. These include a series of independent choices to add one more drainage canal, one more roadway, one more retirement village, and one more well to provide Miami with drinking water. No one chose to reduce the annual surface flow of water into Everglades National Park, to intensify the effects of droughts, or to encourage unnaturally hot and destructive fires. Yet all of these things have happened, and, at this point, it is not clear how the "decision" to degrade the Everglades can be reversed.

Each threatened and endangered species, with a few exceptions, owes its special status to a series of small decisions. Polar bears, key deer, bald eagles,

California condors, Everglades kites, humpback whales, and green turtles have all suffered from the combined effects of single decisions about habitat conversion or overexploitation. In the case of the green turtle, the removal of nesting beaches one by one through development and human intervention has paralleled the decline of green turtle populations. Furthermore, this decline has been accelerated by a multitude of independent decisions by individual fishermen to harvest one more turtle despite their recognized threatened status.

The insidious quality of small decision effects is probably best exemplified by water and air pollution problems. Few cases of cultural eutrophication of lakes are the result of intentional and rational choice. Instead, lakes gradually become more and more eutrophic through the cumulative effects of small decisions: the addition of increasing numbers of domestic sewage and industrial outfalls along with increasing runoff from more and more housing developments, highways, and agricultural fields. Similarly, the gradual decline in the air quality of the Los Angeles basin during the 1940s and 1950s was produced by thousands of small decisions to add one more factory or one more family automobile.

Obviously, Alfred Kahn's observation concerning the net effect of small decisions has great applicability to problems. We could add many more examples to our list, including the decline of prime farmland in the United States, desertification, misuse of groundwater resources, the impact of persistent pesticides, the side effects of single-species management in fisheries and wildlife management, the threat of tropical forest clearing, and the increasing severity of acid precipitation.

Loosing the Chains

While it is easy to recognize this basic problem in the environmental decision-making process, it is not so simple to do anything of a corrective nature. One apparent step would be to strengthen and protect the upper levels of environmental decision-makers (the Department of the Interior, NOAA, EPA, etc.). Unfortunately, these organizations do not always operate with the greatest efficiency, become entangled in their own bureaucratic red tape, and, in the end, leave decisions to the lower levels by default.

Moreover, most of the rewards and pressures within both contemporary political and scientific systems force us toward specific problems and specific solutions, in other words, small decisions. In the political realm, the trend is toward decision-making at lower levels of the system (e.g., the "new

federalism" of Ronald Reagan). Although this may be successful for relatively simple problems, such as building schools, this type of approach offers
little hope for solving complex problems of environmental management.
Unfortunately, it is much easier and politically more feasible for a planner or
politician to make a decision on a single tract of land or a single issue rather
than attempting policy or land-use plans on a large scale.

This pattern of rewards, pressures, and trends is not unique to politics but
also permeates academic science. The majority of scientists are most com-
fortable concentrating upon pieces of problems rather than on an entire
system. In medicine the trend since the time of Louis Pasteur has been toward
single-cause and single-effect medicine ("germ theory"), with modest emphasis on total body responses ("holistic medicine"). Reinforcing this reductionist tendency in science is the coordination of both grant money and
academic tenure with the solution of short-term problems (i.e., small problems).

One key to avoiding the problem of the cumulative effects of small
environmental decisions lies in a holistic view of the world around us.
Scientists, no matter how reductionist their research, should be able to
understand and predict how their specialty fits into whole-system processes.
In addition, we must have at least a few scientists who study whole systems
and help us to avoid the consequences of small decisions. Conversely,
planners and politicians must have a large-scale perspective encompassing
the effects of all their little decisions. Most important of all, environmental
science teachers should include in their courses examples of large-scale
processes and resulting man-induced problems (e.g., the Florida Everglades,
the Colorado River, the Amazon Basin).

Sadly, prospects are not encouraging. Few politicians, planners, or scientists have been trained with, or have developed a truly holistic perspective.
Considering all of the pressures and short-term rewards that guide society
toward simple solutions, it seems safe to assume that the "tyranny of small
decisions" will be an integral part of environmental policy for a long time to
come.

Bridging the Four Major "Gaps" that Threaten Human and Environmental Quality

Eugene P. Odum

*W*hen situations seem impossibly complex, it sometimes helps to back off and take an overview. Often there are common denominators that link together components. If so, then it will pay to assess the whole rather than try to deal with all the pieces or issues as if they were separate entities. I believe we can make a good case for the proposition that the "predicament of humankind" as we approach the 21st century cries out for such a holistic approach.

The litany of problems headlined by the media grows daily. Homeless people in the cities, masses of poor in less-developed countries, drug problems, ozone depletion in the stratosphere, too much ozone at ground level, acid rain, global warming, tropical forest destruction, and on and on. Up to now, society has attempted to deal with such problems on a short-term, one-problem/one-solution basis, leading to what economist Alfred E. Kahn has called "the tyranny of small decisions" (see Kahn, 1966; Odum, 1982). Increasing the heights of smokestacks, a "quick fix" for local smoke pollution, is an example of a small decision that alleviates the local problem but increases a regional problem, namely, acid rain. As long as "small decisions" and "quick fixes" are viewed as solutions, the basic or underlying causes and real solutions are not addressed. In the case of the stack discharges, the real solution involves removing the acid-forming contaminants from the coal or other fuel before it is burned. Even though the technology to do this is

---

Reprinted from *Bridges* 1(3/4):135–141, 1989.

available (see Spencer et al., 1986), doing it is a big decision that is difficult to make because it threatens the short-term economic and political status quo. So we find it less painful to try to treat symptoms rather than diseases, to use a medical analogy.

In many less developed countries, a large-scale "quick fix" is underway that has potentially disastrous consequences for the future. Faced with deep poverty and debt (and in many cases, overpopulation), many nations are exploiting their natural resources in an unsustainable manner in order to generate income in the short term. In South America, the Amazon rain forests are being cut down at a rapid rate despite strong evidence that: (1) crop and pasture land that will replace the forest are neither economically nor ecologically viable, and (2) long-term returns from sustainable harvests of forest products are greater than the income derived from destruction of the forests. A similar situation is accelerating in Indonesia, where hillsides are being cleared of vegetation with little concern for erosion that will soon make the land worthless for anything. (These and other examples are documented in Repetto et al., 1989.) And it is not just in poor countries that these sorts of Faustian bargains are being made; right here in the United States, futures are being jeopardized for quick short-term monetary profits. Before we point our finger elsewhere, we must be aware that North Americans contribute far more global air pollution than South Americans or Southeast Asians due to our much greater energy consumption.

A good way to assess, holistically, current dilemmas is to consider the gaps that must be narrowed if humans and environment as well as nations are to be brought into more harmonious relationship. There are four gaps that need our undivided attention.

The Income Gap between the rich and the poor, both within nations and between industrialized nations (30% of the world's population) and the non-industrialized nations (70%)

In recent times, both the rich and the poor have gotten richer in monetary terms, but the gap has increased at an alarming pace. Between 1950 and 1980, the gap in per-capita income between rich and poor nations increased from $3,617 to $9,648 (see Seligson, 1984). And in the so-called "wealthy" nations, the number of people below the poverty line has increased along with crime and alcohol and drug abuse. In 1988, 50 million people, both urban and rural, were estimated to be below the poverty level in the continental United States.

The Food Gap between the well-fed and the underfed

Again, although yields of food crops have increased worldwide, the gap between rich and poor countries has widened. For example, the gap in grain production between rich and poor countries increased some 2,000 kilograms per hectare (kg/ha) between 1970 and 1980 (see UNFAO, 1985). In 1985, rice yield in Bangladesh averaged only 2,100 kg/ha as compared to 6,100 in Japan. Wheat harvest in Argentina was 1,900 kg/ha as compared to 7,200 in The Netherlands (see Odum, 1989a, Table 2, Chapter 4).

The Green Revolution has helped countries such as India and China feed their populations, but, on the whole, the rich nations have benefitted more than the poor ones. This is because, to get the most out of high-yielding cultivars, expensive subsidies — fertilizers, pesticides, and water — are required that poor nations can ill afford. In the United States and Europe, yields have increased so much that more food is raised on less land, but in the poorer countries increases in food production, often just barely enough to keep up with rapid population growth, have come largely as a result of putting more land (much of it marginal) into cultivation.

The Value Gap between market and non-market goods and services

When it comes to money, human-made goods and services are accorded a very high value, while the goods and services of nature involved in air, water, and soil production and recycling are given little or no monetary value. Thus, an urban acre suitable for skyscrapers is worth millions of dollars, while an acre of forested rural watershed is valued at a few hundred dollars, mostly for its timber. The value of the unharvested watershed to people in the city is unpriced, even though the watershed is vital to life in the city; without water from the watershed there would be no skyscrapers. And the same, of course, for clean air, which is completely unpriced in the marketplace.

As long as the supply of these "free" or "common-property" goods and services is far greater than the demand, then this value gap does not matter. But, as is becoming more evident every day, that is no longer the case. Accordingly, life-supporting non-market goods and services have to be protected by political means or else they have to be incorporated in economic cost accounting. Some very big decisions will be required to bridge the value gap by either or both ways, but doing so will ensure that our vital life-support resources are respected and preserved.

The Education Gap between the literate and the illiterate, between the skilled and the unskilled

Those of us who are educated and read books and journals such as this one tend to be shocked by the statistics that indicate that a large percentage of people are functionally illiterate, not only in poor countries but also right here in our own nation. Many people that we meet in our daily lives are very clever at hiding their reading and writing disabilities! And many of those who may be classed as literate are "innumerate," in that they are unable to deal with numbers, i.e., lack simple arithmetic skills.

The last of the "Club of Rome" books — a much admired but little-acted-on series about global problems — was titled *No Limits to Learning.* (The Club of Rome books are available from Pergamon Press, New York, except for the first volume, *The Limits of Growth* (Meadows et al., 1972), which created storms of protest from political leaders and decision-makers, these books that attempted thoughtful assessments of global problems are little read. I think this is because people in general were not ready to consider hard decisions during the 1970s.) In this treatise, a distinction was made between "micro-education" and "macro-education." The former was defined as education of the individual, one by one, as in traditional schools and colleges. The term "macro-education" was suggested for education of masses of people as a whole. It was concluded that society has not yet developed a technology for this level of education, although television would seem to be a key. How, then, do we make large numbers of people, "the public," as it were, aware of the underlying causes and long-term solutions to current "predicaments"?

For one thing, modern life has become so stressful that people, both rich and poor, have neither time to think beyond the narrow confines of everyday existence nor time to improve their education. For example, a person living in a large city is preoccupied with surviving in an economically competitive environment and has no time to think about where food and water come from. The life-supporting natural and agricultural environments are out of sight and out of mind until there is a water or food shortage! Likewise, the millions of poor people in underdeveloped countries who must eke out a bare existence on a day-to-day basis have no time to consider the long-range consequences of their actions (such as cutting down all the trees in the semiarid regions of Africa in order to have fuel to cook supper).

If these gaps are not narrowed soon and population growth is not controlled to some reasonable degree, then there is great danger of widespread social disorder or destructive revolutions. If the non-market resource capital continues to be depleted, the masses of poor — underfed and uneducated — will see no hope for better life, and they will then vent their frustrations on the affluent minority.

A Common Denominator

To my way of thinking, all of the problems and situations and all the gaps we have been discussing have a common cause in excessive (and unnecessary) waste of assets, both material and human, and in the inability of societies to make decisions in favor of sustainable rather than "boom-or-bust" economic development. If we can increase the efficiency of resource use and reduce waste on a large or global scale, then a major step would have been taken to reduce pollution and bridge gaps. A very large percentage of air and water pollution that is threatening basic life-supporting processes is due to wasted resources: for example, inefficient burning of fossil fuels, dumping into the environment of usable and recyclable materials, or careless handling and disposal of toxic materials when less toxic materials are available for use in manufacturing. *Waste reduction*, **not** *waste disposal*, needs to be the focus for the future, and this involves a complete "about-face" in the way in which societies have approached waste problems in the past.

Let us take agriculture as an example. Recent development of technology for "reduced input conservation tillage" has made it possible to reduce by perhaps half the fertilizers, pesticides, and other agricultural chemicals now used at no sacrifice to crop yield. In fact, yields would be more sustainable with less plowing and reduced use of toxic chemicals since soils would be improved rather than poisoned or used up (see Gebhart et al., 1987). If such technology is adopted in the rich countries and if poor countries skip the wasteful phase and go directly from their traditional procedures to efficient reduced-input practices, then the following could result: (1) nonpoint water pollution that results from excessive use of environmentally damaging poisons would be reduced; (2) pesticide and other dangerous residues in food would be reduced; (3) profits for the farmer would be increased as the cost of producing a unit of crop (bushel of corn, for example) is reduced; and (4) the food and income gaps would be narrowed. Comparable scenarios could be outlined for the management of power plants, industries, and so on. (For more on this theme, see Odum, 1989b.)

In other words, if the United States and other rich nations can "power down" by reducing waste without reducing the quality of life for the individual, then we become believable role models for the poorer nations and will encourage them to undertake sustainable development without destroying future environmental and human assets. What we are proposing here is bridging dangerous gaps through international cooperation for mutual benefit, not a utopian and impractical "share-the-wealth" scenario. Nature provides us good clues on how to power down and continue to prosper in a world of limited resources. Coral reefs in the South Pacific, for example, are very prosperous natural cities, productive and crammed with life, despite the fact that the surrounding oceans are nutrient deserts. Producers and consumers (plants and animals) are so closely coupled and nutrients so efficiently recycled that dependence on imports is minimal.

It would also help if we could somehow heed the common-sense proverbs that represent human experience acquired over the ages — for example, "don't put all your eggs in one basket," or "an ounce of prevention is worth a pound of cure." In our current pursuit of short-term wealth, we tend to lose sight of these truisms. Two familiar sayings that are especially appropriate for this commentary are: "haste makes waste" and "waste not, want not."

One of the problems of undertaking major changes in direction is that inevitably there will be transitional economic costs. For example, converting from conventional to conservation tillage, or adding equipment to clean the fuel to be burned in a power plant, require investment in new equipment and procedures. Governments can speed desirable changes by underwriting or subsidizing these transition costs, but they are not likely to do so until large numbers of people are convinced that long-term benefits will be forthcoming. If it is true that people and nations move ahead only when there is a common vision that motivates, then a vision of a better world through waste reduction could very well unite the political left and right (another "gap" of great concern!), thus combining the individual and the public good (see Hawkins et al., 1982).

Biotechnology Presents a Challenge to the Campus

Eugene P. Odum

S cholars in a wide range of disciplines should become involved with biotechnology, not just geneticists and molecular biologists. No matter how specialized a new technology seems in the beginning, successes and failures ultimately depend as much on social, economic, legal, and environmental aspects as on advances in the technology itself. There have been several major "technological revolutions" in my lifetime, and I, along with students and colleagues, have been actively involved in research on several of these. In retrospect, expectations and benefits tended to be exaggerated in the beginning, while problems were often not anticipated and costs were underestimated.

Judging from these past experiences, I believe that real long-term benefits to humanity come only after a wide variety of disciplines become involved and questions of private versus public values are resolved. A brief review of emerging technologies associated with pesticides and atomic energy will serve to illustrate and help put biotechnology, as the newest technological imperative, in perspective. We all remember the exaggerated claims made for DDT and related broad-spectrum persistent insecticides when they first hit the market. No more problems with insects (they said!), and feeding the

First published in the University of Georgia newsletter, *COLUMNS* **14**(16), 1987.

world with pest-free crops would now be a snap. So effective was the propaganda and so lucrative the industrial profits that whole entomology departments became preoccupied with spray-gun chemistry, i.e., no need to study bugs, just spray on the magic potions! Of course, much good has come from pesticide technology, especially as the "revolution" has matured and focused on species-specific approaches to pest control. But unanticipated deleterious side effects, especially on beneficial insects and birds, became so severe and lasting in time that DDT and some of its relatives had to be banned to protect environmental and human health. I might point out that those persons who first documented the deleterious effects and urged caution on mass spraying were by and large neither chemists nor entomologists (Rachel Carson, for example).

At the first international conference on the peaceful uses of atomic energy in 1955, which I attended as an official U.S. delegate, the hyperbole was unbelievable. The chairman of the conference, a scientist from India, predicted in a keynote address that electricity generated by atomic energy would be so cheap by 1975 that utilities would not bother to meter it (just charge a flat fee and let everyone use as much as they wanted!). He also said that all nations would be equal in energy supply in the atomic age since "the atom is everywhere!" When some of us asked about waste disposal, we were assured that this would be but a minor engineering problem.

When I returned from the conference, I was both so excited and worried about the coming atomic age that I took a year's leave of absence in order to study atomic energy and visit places where the first research on environmental effects was in progress. This experience convinced me that ecologists should become involved in research on both positive and negative aspects. Accordingly, we recruited new staff and students and established an Institute of Radiation Ecology (which later evolved into the Institute of Ecology). As a result, the University of Georgia became a pioneer in the development of a new interdisciplinary field. At first, atomic physicists and engineers resented ecologists "horning in on their field," but exchange of ideas and skills soon proved to be mutually beneficial. Whatever may be the future of atomic energy, I can say that ecology has become a much stronger discipline as a result of the cross-fertilization.

Now we come to biotechnology. The chemists and physicists have had their time in the public limelight, and now it is the biologists' turn. But, by and large, the market orientation of biotechnology is a new experience for academic biologists, who are more accustomed to federally supported basic research in the public interest than to industry-supported research for profit.

Indeed, the rapidity with which industry has moved to exploit recombinant DNA, monoclonal antibodies, and other molecular biological breakthroughs makes biotechnology unique among the new technologies that have emerged in this century. University administrators are on the horns of a dilemma, since, on the one hand, they welcome the large funding the states and industries seem willing to invest, but, on the other hand, they are concerned that the university might become too much a captive of the military–industrial complex. An article by Leon Wofsy in the *Journal of Higher Education* (**57**(5):477–492, 1986) discusses this dilemma in detail. Carl Kaysen, a professor of political economy at MIT, writing in the *Bulletin of the American Academy of Arts and Sciences* (**39**(7):19–32, 1986), predicts that there will only be a modest increase in industry's funding of university research and that support for basic research will have to continue to come from public and foundation sources.

Funding problems aside, we can expect many good things for health and medicine, agriculture, and economic development (including benefits for poor countries) to result from biotech research. But we can also expect unanticipated costs and problems along with environmental, ethical, social, and legal questions that will need to be researched. To ensure that the positives outweigh the negatives, the entire spectrum of disciplines from the humanities to the sciences should get involved, as I suggested at the beginning of this commentary. Since the University of Georgia is somewhat of a "Johnny-come-lately" to biotech (Harvard and MIT had already negotiated multimillion-dollar contracts with industry in 1974), a broad approach based on the diversity of talent that is to be found only in a major state university would seem to be our best strategy. This would also counter the current overspecialization that hampers the university's efforts to deal with general education and real-world problems that, by and large, require an interdisciplinary effort. Those of us in the Institute of Ecology are already joining the bandwagon. We have established a working group in environmental biotechnology that builds on our existing strengths in microbial ecology, agroecology, and population genetics. Four candidates have been interviewed for a new position in environmental biotech, and a lecture series has been established by endowment funds.

I would suggest that the vice president for research establish a general research fund and invite proposals from the faculty at large. Understandably, many faculty will be hesitant to get involved because of a lack of expertise in subjects related to genetic engineering, but professors and students can be quick learners (as we were in the case of atomic energy). Joining together with knowledgeable persons for team research and education is especially desirable at this time. Remember, there are very few "experts" in the beginning of a scientific revolution. The challenge is there for anyone concerned with the implications who is willing to try breaking out of his or her narrow specialty.

Source Reduction, Input Management, and Dual Capitalism

Eugene P. Odum

*R*ecently, a friend who teaches in junior high asked me to visit with her class and discuss what we learn from ecology about understanding and dealing with human predicaments, the subject of my new book in preparation, entitled "Ecological Vignettes." Since I am using cartoons, along with simple charts, to illustrate each vignette, I was eager to see how kids would respond to the use of cartoons to introduce important environmental problems.

One cartoon that I projected on the screen by Sidney Harris shows a large tanker truck with "CAUTION: HAZARDOUS MATERIAL" written in large, black letters on the side (see Chapter 8, Figure 8.1). The driver is saying to a passenger, "Didn't you know? We just drive around. This is a mobile toxic waste dump." After everybody laughed, I asked, "Is there a basic message here?" Several hands went up with answers to the effect that "there is no longer any place we can dump toxic wastes." Then I asked, "Then, what are we going to have to do?" There was a longer pause, but a student in the back of the room suggested, "We will just have to stop producing the stuff." So, there you have it from the mouth of a child: SOURCE REDUCTION AS THE ULTIMATE SOLUTION TO POLLUTION!

Published in the *Journal for Cleaner Production*, 1997, in press.

Source reduction is part of what I have called "input management" (Odum, 1989), which involves focusing on inputs to production systems (industries, power plants, agriculture, and so on) so as to increase efficiencies and reduce or eliminate environmentally damaging materials at their onset. The point is that this is the only way to deal with nonpoint pollutants since humans can do very little once the pollutants are output and widely dispersed into our life-support environment. One can only hope that nature can absorb them without too much damage, or without large future costs of restoring damaged ecosystems (as with the Sudbury smelter-made desert; see Gunn, 1995).

It is encouraging that pollution reduction success stories are to be found in the pages of the journal *Pollution Prevention Review*. For example, Kirsch et al. (1993) describe how four different manufacturers reduced very troublesome painting wastes by either separating and recycling the solvents or, better still, replacing solvent-based painting with electrostatic powder coating. But such efforts are not going to be enough in the future. We need to develop an ecological economics in which the non-market goods and services of the biosphere that maintain the quality of air, water, and soil (the life-support systems) are accorded value equal to that of human-made market goods and services on which current market capitalism is based. It should be feasible with an appropriate regulatory incentive motivation to develop such a *Dual Capitalism* over a period of time. For example, a new business or industry would not only consider the market possibilities for the new product or service but also ways to produce the product or service with efficient use of resources, with as much recycling as possible and with as little pollution as possible (lots of opportunity for new technology here!). In this way, natural capital and human or market capital can be given equal attention. Also, it will be important to internalize the cost of waste reduction and management so that the consumer rather than the taxpayer pays for the increasing cost of maintaining quality of life in a world of increasing human population and declining per-capita resources.

Earth Stewardship

Eugene P. Odum

W e are hearing a lot these days about "stewardship of the earth," an entreaty that hardly anyone is against, but many are not strongly for it. Few, if any, politicians actually come out and say they don't approve of it, yet many hesitate to introduce legislation to promote it. And we don't see people on the street with signs exclaiming "down with steward-ship," yet we see many people on the street acting in a very unsteward-like manner. As with many strategies that seem desirable in the abstract, people want to know what it means to individuals and human communities to practice earth stewardship; how will it affect lifestyles and taxes, and why is it important? It is a little bit like N-l-M-B-Y; a good thing and we ought to do it, but not necessarily in my backyard!

The word "steward" comes from roots meaning "keeper of the house." We use the word widely for a person who manages or takes care of property, money, a ship, restaurant, airplane (flight attendant), and so on. We expect, of course, a steward to act in the best interest of both the owner and the public; that is, a proper steward is ethical and does not cheat. And, in general, society enacts severe penalties if a trusted steward does otherwise when it comes to private or governmental property. The question now is: has the time come in history when developing a stewardship relationship between hu-mans and the earth's life-supporting common property (air, water, soil, biodiversity, etc.) is necessary for our continued well-being or even our survival? Is the time coming very soon when it will not only be unethical to mistreat Mother Earth but unlawful as well? I believe we can make a good

Reprinted from *Earth Ethics* **3**(3):11–12, 1992.

case for answering these questions in the affirmative. Let me now present a biblical and an ecological analogy to illustrate why and how it is in our best interest to become better "keepers of the house."

Messages from the Scriptures

Early on in the scriptures, we are told to multiply and take dominion over the earth, but later on in these same scriptures we are also told to be stewards and take good care of the earth. A reasoned interpretation is that these messages are not contradictory, nor a matter of right and wrong, but demonstrate a sequence in time; that is, whether we focus on "dominionship" or "steward-ship" depends on the stage of development of human society. In the early or pioneer stages of civilization, exploitation of the earth's resources (clear the land for agriculture, mine the earth for materials and energy, and so on) and a high birth rate are necessary for survival. But as society matures and becomes increasingly crowded and technologically complex, there is less need for large families and, more important, various earth limitations force us to turn to stewardship in order to avoid destroying our life-supporting "house," the biosphere.

The transition from youth to maturity, like adolescence in the individual, is a difficult and dangerous time. If we are to survive the transition and continue to prosper, we now have to be willing to divert more ingenuity and more of our tax revenues or other wealth to servicing, that is, maintaining the functional health of the natural environment as well as the quality of the human-built environment. Accordingly, efforts now underway to develop infrastructures for recycling, energy conservation and diversification, natural area preservation, and reasonable family planning need to be accelerated. And cooperation among nations in regard to these needs is desperately needed. Survival of our nation involves a lot more than just confronting "evil empires" (Russia and Iraq).

Let me emphasize that stewardship strategies are not "anti-growth," as many vested interests would have us believe. We all know deep down that there are limits to growth in the size of an individual, a community, or the population of the earth, but as far we know there is no limit to growth in quality. The youth → maturity transition involves a gradual shifting from quantitative to qualitative growth.

For example, our city of Athens is at a point in its history where management is beginning to consider how to become a better city and not just a bigger one. Our university has already set such a course. It is encouraging that the newly elected government for the combined Athens–Clarke County area has come out for an emphasis on quality of life. It is also encouraging that the State of Georgia (and many other states as well) have recently enacted "growth strategies" legislation intended to promote local and regional planning.

But don't expect change overnight. The "bigger is always better" philosophy is strongly entrenched in America, and there are many economic, political, and tax procedures and traditions that actually promote rapid haphazard growth (build now and worry about the consequences latter). Many of these excessive growth-promoting procedures and attitudes will have to be modified if any kind of long-range planning is to be successful. And this won't happen until our political representatives are convinced that a majority of us are in favor of stewardship policies.

The Parasite–Host Model

Since humans are ecologically speaking "heterotrophs," we depend on other organisms for food, and we depend on the natural environment in general for air, water, and many essential materials. Large cities with their huge demands for energy and resources require very large areas of non-urban environment to support them, and it is this life-supporting environment that we are concerned about as the world becomes increasing urban. In a very real sense, humanity is parasitic on its host the biosphere. To speak of ourselves or our city as parasites is not to belittle but to be realistic.

Ecological research on parasite–host interactions in nature are revealing that natural selection operates to promote reciprocal adaptations so that the parasite does not exterminate its host, and, thereby, itself. Given time and freedom from excessive outside disturbance, hosts develop resistance and parasites reduce virulence. For example, when the myxoma virus was first introduced in Australia to control rabbits, the parasite was very virulent and killed the rabbit host very quickly, but, believe it or not, the virulent strain gradually replaced itself with a much less virulent strain, so that both parasite and host survive to this day. We even see in some parasite–host systems what we call "reward feedback," where the parasite actually promotes the welfare of the host, just as we promote the welfare and continued survival of cultivated plants and domestic animals on which we depend. Just recently, there was an article in *Science* magazine that described how a protozoan parasite of a rodent alternates the degree of virulence according to season. During the breeding season of the rodent, when there are lots of individuals, the parasite is virulent, but during the nonbreeding season the parasite becomes avirulent, allowing plenty of hosts to survive to the next breeding season. In nature, the greatest danger of extinction of host and parasite is early in the interaction before there is time for adjustment, or when there is external disturbance that interferes with the "co-evolutionary" process.

The point of all this is that we can learn a lot from ecological studies of development and evolution in nature that provide clues to help us deal with analogous human predicaments. We certainly need to learn how to be a prudent parasite that maintains sustainable parasite–host relationships, rather than continuing to act like a malignant cancer that would ultimately destroy our biospheric host.

The Role of the Professional Environmental Designer

For many years now environmental designers have worked hard to make plans for a better environment, first on a small scale — individual yards, estates, small public parks — and now on the larger scale of whole towns and large landscapes. It is often very discouraging work, as plans are ignored and plans and zoning so easily overturned by short-term economic incentives. My message to students of environmental design is to hang in there tough, because the cards on the table or the writing on the wall indicates that your work will be much more appreciated in the near future. And, as I have indicated, you need to keep abreast of new developments in the discipline of ecology!

Ecology:
The Common-Sense
Approach

Eugene P. Odum

*D*uring the Industrial Age, our confidence in the omnipotence of science and technology has led us increasingly to divert from the path that would have been dictated by the common sense embodied in our traditional culture and so admirably reflected in its proverbs.

Ecology, seen as an approach, rather than as a scientific discipline, provides the rationale for a return to common sense. It is in terms of this approach that we should consider the basic problems that our society faces today.

Sometimes the way to deal with complexity is to search for overriding simplicity. When situations appear hopelessly complicated, as do, for example, the energy and environmental dilemmas that we face today, then it pays to back off a bit and take a broad overview. When the traditional reductionist approach of piecemeal analysis fails to achieve long-term solutions, then it is time to consider a more holistic approach in which interactions of the pieces, and ways of dealing with the situation as a whole, are also considered. This is what ecology as an "emerging new integrative discipline" is all about.

Surprisingly often, when it comes to comprehending and acting on crisis situations, the overriding simplicity turns out to be good old-fashioned

---

Reprinted from *The Ecologist* 7(7):250–253, 1977.

common sense, or "horse sense" as it is sometimes called. For example, when faced with a shortage of something, the common-sense response would be to conserve and reduce waste first, then consider the alternatives for maintaining or increasing supply. Yet, too often of late we go for the technological quick fix that promises to postpone the bother of readjusting, however slightly, our lifestyles. In times of rapid technological growth, the momentum produced by massive energy and material consumption is so strong that the negative feedback signals that warn of "too much of a good thing" are often not heeded until a real crisis occurs. It is also human nature to put off for tomorrow what should be done today. But we cannot ignore for long the common-sense wisdom that has sustained mankind for untold centuries, because to do so invites a Faustian bargain in which the promise of today's technological magic could be all too easily traded for a hell on earth tomorrow.

It is my theme that the natural laws that underlie the principles of ecology can be applied in a common-sense manner that might appeal to people both rich and poor (and in both "developed" and "underdeveloped" nations), who are not inclined to listen to these strident environmentalists that seek to accentuate rather than ameliorate the conflicts between man and nature. While detailed knowledge and extensive training and experience are necessary if one aspires to be a truly dedicated and objective professional ecologist, the basic principles of ecology as they relate to our everyday lives are not all that difficult to comprehend because by and large they are already part of our heritage, even if we are inclined to forget about them temporarily as we enjoy the fruits of material growth. To illustrate, let us consider what might be included in a layman's guide to ecology. Let us avoid the technical jargon that is appropriate to a college text or other professional treatise and see if we can communicate the basic concepts of a "holistic" ecology in everyday language.

The Houses of Man

Since the word "ecology" is derived from a Greek root meaning "house," it is appropriate to think of ecology as the study of our environmental house. Actually, there are two "houses" to be considered when it comes to coping with our environmental problems, namely, the house that man has built and the house of nature. Since these entities are actually open systems, not closed-in units as the word "house" might imply, and since energy is a common denominator, we in ecology often designate the man-made house as the *fuel-powered urban-industrial system* and nature's house as the *solar-powered natural ecosystem*. It is also useful to think of agriculture as an intermediate system that links together the houses of man and nature. Since we now use high-quality energy as well as human and animal labor to augment the abundant but low-quality sun energy to produce food, we can conveniently designate farming operations as a *fuel-subsidized, solar-powered agroecosystem*. As more and more auxiliary fuel energy is used to power machinery, to make fertilizers and pesticides, to irrigate, and to transport and process food, the agroecosystem comes more and more to be like the industrial system in terms of demands and impacts on the natural environment.

Because technological achievements have made us less and less directly dependent on the natural environment for our daily needs, modern man has come to think and act more and more as if the man-made house was independent of the house of nature. Our economic systems of whatever political stripe tend to place a very high value on man's works and almost no value on the works of nature, which are taken for granted, or even downgraded in the sense that "familiarity breeds contempt," as the old saw goes. The great paradox is that from a technological viewpoint the success of industrialized nations has been accomplished by ingenious *uncoupling* of man and nature and by exploitation of the finite, and now declining, nationally produced fossil fuels stored in the earth. Yet man-made systems remain as they have always been, completely dependent on the natural environment, not only for energy and materials, but also for the even more vital support of life processes. As the intensity of man-made developments increases, the impact on the life-support system becomes increasingly critical with air, water, and food increasingly contaminated or in short supply or both. Fortunately there is an increasing awareness worldwide that the energy, food, water, and other "crises" that periodically come to public attention all have an environmental basis, and that solutions require a new or reoriented technology as well as restraints on population growth. It would seem obvious that serious attention needs to be given to the recoupling of man and nature so that the total environmental house can be preserved, and, hopefully, the quality of life therein improved. Thus, the "new" ecology that I preach is dedicated to the common-sense notion of man and nature as a coupled system.

The Whole Is More than a Sum of Parts

When parts function together to produce a whole (or a "system" in more formal language), new properties may appear that are not features of the parts, as each might operate separately. For example, water has unique characteristics that are neither the same as, nor the sum of, the properties of hydrogen and oxygen, its component parts. So it is with the systems of man and nature. Much can be learned by dissecting and studying the component parts of an organism or a forest or a city, but it is necessary to examine the intact or "in situ" or "in toto" system in order to have a full understanding of the situation in question.

In everyday life there are many expressions and sayings that indicate a general acceptance of what we might call the holistic doctrine. Thus, when someone takes an overly narrow view of something we may accuse that person of "not seeing the forest for the trees" or of having "tunnel vision." The trouble is that as a society we do not practice what human wisdom preaches. As we have already noted, science and technology, especially in the last 50 years, have moved in the opposite direction, that is, towards increasing specialization, more subdivisions, more one-problem/one-solutions, and so on. The sad fact is that our really big and important problems, such as those associated with energy, food, water, cities, and population growth, cannot be solved, or even coped with, on the basis of piecemeal study no matter how sophisticated or technically advanced are the methods employed. Thus, "the whole is more than the sum of the parts" is an appropriate common-sense translation of what we would call the "ecosystem principle" in an ecology text.

Haste Makes Waste

This familiar admonition makes a good heading for a chapter on "energy in ecosystems," since it expresses an important aspect of the entropy law, also known as the second law of thermodynamics, one of the most important natural laws. The success of any system, whether man-made or natural, depends not only on the quantity and quality of its energy source but also on how efficiently the source is converted into useful work capable of maintaining the system as a whole. As energy is converted from one form to another to accomplish a useful function or transformation, the quantity is reduced by an inevitable heat loss, but the quality of that which is passed on may be increased. Thus, in the well-known food chain conversions it takes between

500 and 1000 units (calories, for example) of low-quality sun energy to make 10 units of higher-quality plant material, which in turn can be converted into 1–2 units of still higher-quality meat.

It is important to remember that the flow of energy is always one way. Once useful work has been accomplished in a conversion, that energy is no longer available. Energy can be stored, but it cannot be recycled as can such materials as water or iron. The food you ate and the fuel you burned today is gone forever and must be replaced by new food and new fuel tomorrow. Equally important, one cannot be *both fast and efficient* at the same time when it comes to energy conversion. We all know that driving very fast reduces the distance we can go on a given amount of fuel. Haste does indeed increase waste; fast conversions mean less work accomplished for a given amount of energy and also more heat and other waste products generated.

When easily convertible energy is plentiful, then both man and nature tend to hasten and make waste, which in turn requires additional energy to cope with the disorder created by the growth in size and by the waste products produced. In theory, negative feedback then acts to slow down the haste and increase efficiency. We do observe that in natural ecosystems growth slows down, energy is used more efficiently, and stores of high-quality energy are established in the biomass as the ecosystem develops from a pioneer stage to maturity. The same things happen if the inflow of energy or its quality is reduced for whatever reason. The common-sense notion of "save for a rainy day" becomes appropriate when saturation levels of use are approached since energy and resources are always subject to periodic fluctuations in the real world.

As already noted, man should respond to an energy crisis in the same general way as does nature, but energy conservation as a public policy does not have the appeal that the search for glamorous new sources provides. Unfortunately, finding new sources does not necessarily resolve an energy crisis if a lot of the newfound energy has to be used to develop and maintain the new flow and to deal with new and perhaps more toxic waste products. For example, fusion atomic energy might not prove to be the bonanza we expect since much energy will be needed to cool down the reaction from millions of degrees to a usable level. So far in fusion research the break-even point, where as much energy is produced as is required to produce it, has not been achieved even on a small scale.

Not only do natural laws rule against having speed and efficiency at the same time, but they also make it difficult to have high quality and large quantity simultaneously. Increasing the quantity of resources increases the potential for rapid growth, but such growth may come at the expense of the quality of the individual and/or the quality of life for the individual. In the extreme, fast growth can become disorderly like cancer and threaten survival of life itself. The eutrophication (enrichment by pollutants) of natural lakes provides an illustration of the quantity–quality dilemma. When nutrients from sewage are put into the lake, the number of organisms and the rate of organic production increases but "weed-type" organisms such as small "scummy" algae and "trash" fish replace the diatoms, attractive water plants, and game fish. If enrichment is intensified, more and more kinds of organisms are eliminated even as those that remain multiply like the out-of-control cells in a cancerous organ. One cannot be certain that the discovery of a new unlimited and cheap energy source, granted that it's possible, would really be a boon for humankind. It might just be "too much of a good thing" that would convert the world into one big overpopulated cesspool, an undesirable "whole earth" if ever there was one!

All in all, then, the judicious solution to the energy, food, water, or what-have-you crises is to cut down on haste in order to reduce waste, increase efficiency and buy time to improve the quality of human life. At the same time, without undue haste, we can look into our options for adjusting supply and demand. To act on such common-sense judgment requires not only science and technology dedicated to such goals, but more difficult to achieve reordered political and economic objectives, which today are much too strongly geared to promote growth and waste, or quantity "uber alles."

Don't Put All Your Eggs in One Basket: Variety is the Spice of Life

These two common sayings would make good headings for a chapter in "community ecology" including the general subject of "diversity." Few of us would disagree that it is unwise, and usually downright foolhardy, to put all one's money, or whatever, into only one venture. Naturalists for centuries have marveled at the diversity of life in natural systems, and modern ecologists have generally agreed that there is efficiency and safety in diversity, although they are not sure just why and how diversification evolves. The idea is that a diverse ecosystem is better able to use the energy and resources available and better able to resist adversity. However, it is also clear that diversity has an energy cost of its own so that one can have too much as well as too little variety. Right now our concern in human society is with too little. Industrial societies have often thrived on a short-term basis by putting all their eggs in one basket. Thus, in the United States we put most of our transportation eggs into the automobile basket, a lot of the energy eggs in the oil basket, too much of our hairspray in aerosol cans, and so on. Also, more and more we concentrate on one or very few kinds of grain, or species of trees, for the food and forest baskets. All the while, we seem fully aware that this sort of strategy invites the overshoot, the boom and the bust, as it were. One reason we do it is that high profits and rapid growth come when we concentrate on promoting single products. We assume that when the diminishing returns set in we can quickly and easily shift to another basket. But what if we do not have another basket ready when the one we have been using breaks, or what if the eggs we lost have not been paid for? Then there will be serious "transition losses," including perhaps economic depressions and social disorders as we struggle to recoup losses and organize another basket. If we can only heed the common-sense warning in these matters and recognize and act on the premise that variety is not only the spice of life "quality factor" but also a valuable stabilizing factor, then we should be able to devise the means to promote diversification. As already noted, there would be a cost to such a strategy, and we would have to be convinced that this cost is less than the transition losses inherent in the boom-and-bust model. The evolutionary success of diversification in nature would seem to indicate that this is indeed that case.

The Bottom Line

My theme has been that common-sense notions representing human wisdom of the ages provide a basis for seeking holistic solutions of problems of energy and environment that appear hopelessly complicated when viewed piecemeal. Also, I cited several examples to show how the principles of ecology can be expressed in terms of this common-sense wisdom. When we do take an overview of problems related directly or indirectly to environment, it becomes clear that the time has come to recouple the two "houses" of man, the man-made urban-industrial system and the natural environment life-support system. During the Industrial Age, these two vital parts of our total existence have become too far separated in our minds and actions, leading to dangerous inequities of value and performance. To recouple man and nature into a more harmonious whole requires that science and technology be integrated with reordered social, economic, and political goals — a most difficult task. It is tempting to wish for a benevolent dictator who could act for the good of the whole so as to prevent, or at least blunt, the overshoot that comes with going too fast too far. But we are immediately reminded of another wise saying that "power corrupts"! Benevolent dictators do not remain benevolent for very long. Being human, they are likely to abuse their power to the detriment of basic human rights. The best we can hope for is a massive educational effort at all levels of the media so that an informed public can select leaders who see the whole as well as the parts; and who, we might add, can be recalled if they abuse their political power.

Literature Cited

Essay 1:
Harmony Between Man and Nature: An Ecological View
Eugene P. Odum

Cook, Earl (1971). The flow of energy in an industrial society. *Sci. Am.* **224**(3):134–144.

Fiebleman, J.K. (1954). Theory of integrative levels. *Brit. Jour. Phil. Sci.* **5**:59–66.

Forrester, Jay W. (1971). *World Dynamics*. Wright-Allen Press, Cambridge, MA.

Gelfant, Seymour, and J.G. Smith Jr. (1972). Aging: noncycling cells an explanation. *Science* **178**:357–361.

Gordon, D.M. (1995). *American Scientist*, **83**:50.

Heller, Alfred (1971). *The California Tomorrow Plan*. William Kaufman Inc. Las Altos, CA.

Meadows, D.H., D.L. Meadows, J. Randers, and W.H. Behrens III (1972). *The Limits of Growth*. Universe Books, New York.

Miller, James G. (1965). The organization of life. *Perspectives Biol. Med.* **9**:107–125.

Odum, Eugene P. (1969). The strategy of ecosystem development. *Science* **164**:262–270.

_____. (1971). *Fundamentals of Ecology*, 3rd ed. W.B. Saunders, Philadelphia.

Odum, Eugene P., and Howard T. Odum (1972). Natural areas as necessary components of man's total environment. *Trans. 37th N.A. Wildl. Nat. Res. Conf.*, pp. 178–189. Wildlife Management Institute, Washington, DC.

Schurr, Sam H., ed. (1972). *Energy, Economic Growth, and the Environment*. Johns Hopkins University Press, Baltimore.

Essay 3:
The World's Most Polymorphic Species: Carrying
Capacity Transgressed Two Ways
William R. Catton Jr.

British Museum of Natural History Staff (1981). *Origin of Species.* Cambridge University Press, Cambridge.

Calabresi, G., and P. Bobbitt (1978). *Tragic Choices.* W.W. Norton, New York.

Catton Jr., W.R. (1985). Emile who and the division of what? *Sociological Perspectives* **28**:251–280.

_____. (1986). *Homo colossus* and the technological turn-around. *Sociological Spectrum* **6**:121–147.

Colinvaux, P.A. (1973). *Introduction to Ecology.* John Wiley & Sons, New York.

Costanza, R. (1987). Social traps and environmental policy. *BioScience* **37**:407–412.

Duncan, O.D. (1959). Human ecology and population studies. In: *The Study of Population* (P.M. Hauser and O.D. Duncan, eds.), pp. 678–716. University of Chicago Press.

_____. (1961). From social system to ecosystem. *Sociological Inquiry* **31**:140–149.

Durkheim, E. (1933). *The Division of Labor in Society* (G. Simpson, trans., original French ed., 1893). Macmillan, New York.

Emlen, J.M. (1984). *Population Biology: The Coevolution of Population Dynamics and Behavior.* Macmillan, New York.

Ford, E.B. (1955). Polymorphism and taxonomy. *Heredity* **9**:255–264.

Freese, L. (1985). Social traps and dilemmas: where social psychology meets human ecology. Paper presented at the annual meeting of the American Sociological Association. Washington State University, Pullman, WA.

Gaskin, D.E. (1982). *The Ecology of Whales and Dolphins.* Heinemann, London.

Hardin, G. (1968). The tragedy of the commons. *Science* **162**:1243–1248.

_____. (1986). Cultural carrying capacity: a biological approach to human problems. *BioScience* **36**:599–606.

Hawley, A.H. (1973). Ecology and population. *Science* **179**:1196–1201.

Kahn, H., W. Brown, and L. Martel. (1976). *The Next 200 Years: A Scenario for America and the World.* William Morrow, New York.

Kleiber, M. (1947). Body size and metabolic rate. *Physiol. Rev.* **27**:511–541.

McHale, J., and M.C. McHale (1976). *Human Requirements, Supply Levels and Outer Bounds: A Framework for Thinking about the Planetary Bargain.* Aspen Institute for Humanistic Studies, Publishing Program Office, Palo Alto, CA.

Minasian, S.M., K.C. Balcomb III, and L. Foster (1984). *The World's Whales: The Complete Illustrated Guide.* Smithsonian Books, Washington, DC.

Odum, E.P. (1969). The strategy of ecosystem development. *Science* **164**:262–270.

Office of Technology Assessment (OTA) (1985). *Technology, Public Policy, and the Changing Structure of American Agriculture: A Special Report for the 1985 Farm Bill.* U.S. Congress, Office of Technology Assessment, OTA-F 272: March.

Park, R.E. (1936). Human ecology. *Am. J. Sociol.* **42**:1–15.

Pratt, W.E. (1952). Toward a philosophy of oil finding. *Bull. Am. Assoc. Petroleum Geologists* **36**:2231–2236.

Remmert, H. (1980). *Arctic Animal Ecology.* Springer-Verlag, Berlin.

Shelby, B., and T. Heberlein (1984). A conceptual framework for carrying capacity determination. *Leisure Sciences* **6**:433–451.

Simon, J.L., and H. Kahn (1984). *The Resourceful Earth: A Response to Global 2000.* Basil Blackwell, New York.

Topoff, H., ed. (1981). *Animal Societies and Evolution.* W.H. Freeman, San Francisco.

von Liebig, J. (1842). *Chemistry in Its Application to Agriculture and Physiology.* Taylor & Walton, London.

Wagar, J.A. (1964). The carrying capacity of wild lands for recreation. *For. Sci.*, Monogr. No. 7, pp. 1–24.

Welty, J.C. (1982). *The Life of Birds*, 3rd ed. Saunders College Publishing, Philadelphia.

Whittaker, R.H. (1975). *Communities and Ecosystems*, 2nd ed. Macmillan, New York.

Winner, L. (1986). *The Whale and the Reactor: A Search for Limits in an Age of High Technology.* University of Chicago Press.

Essay 8:
Ecosystem Management: A New Venture for Humankind
Eugene P. Odum

Adkisson, P.L., Niles, G.A., Walker, J.K., Bird, L.S., and Scott, H.B. (1982). Controlling cotton's insect pests: a new system. *Science* **216**:19–22.

Allen, T.F.H., and Starr, T.B. (1982). *Hierarchy: Perspectives for Ecological Complexity*. University of Chicago Press.

Johnson, W. Carter, and Sharpe, D.M. (1976). An analysis of forest dynamics in the northern Georgia Piedmont. *Forest Sci.*, **22**:307–322.

Jordan, Carl F. (1982). Amazon rain forest. *Am. Scient.* **79**:396–401.

Kahn, Alfred E. (1966). The tyranny of small decisions: market failures, imperfections, and limits of economics. *Kyklos* **19**:23–47.

Livingston, R.J. (1980). *Understanding Marine Ecosystems in the Gulf of Mexico*. U.S. Man and the Biosphere Program, Report No. 2, Washington, DC.

Lovelock, James E. (1979). *Gaia: A New Look at Life on Earth*. Oxford University Press, New York.

Mayr, Ernst (1982). *The Growth of Biological Thought*. Belknap Press of Harvard University Press, Cambridge.

Muscatine, L.C., and Porter, James (1977). Reef corals: Mutualistic symbiosis adapted to nutrient-poor environments. *BioScience* **27**:454–456.

Norman, Richard W. van (1963). *Experimental Biology*. Prentice-Hall, Englewood Cliffs, NJ.

Odum, Eugene P. (1983). *Basic Ecology*. Saunders College Publishing, Philadelphia.

Odum, Eugene P., and Biever, L. (1984). Resource quality, mutualism, and energy partitioning in food-chains. *Am. Nat.* **124**:360–376.

Odum, William E. (1982). Environmental degradation and the tyranny of small decisions. *BioScience* **32**:728–729.

Patten, Bernard C. (1978). The system approach to the concept of environment. *Ohio J. Sci.* **78**:206–222.

Patten, Bernard C., and Odum, E.P. (1981). The cybernetic nature of ecosystems. *Am. Nat.* **118**:886–895.

Simon, Herbert A. (1973). The organization of complex systems. In: *Hierarchy: the Challenge of Complex Systems* (H.H. Pattee, ed.). pp. 1–27. George Braziller, New York.

Waring, R.W., ed. (1980). *Forests: Fresh Perspectives from Ecosystem Analysis*. Oregon State University Press, Corvallis.

White, Lynn (1980). The ecology of our sciences. *Science* **208**:72–76.

Essay 9:
The Watershed as an Ecological Unit
Eugene P. Odum

Hibbert, A.R. (1967). Forest treatment effects on water yield. In: *Symposium on Forest Hydrology* (W.E. Sopper and H.W. Lull, ed.), pp. 527–543. Pergamon Press, New York.

Murphy, E.F. (1967). *Governing Nature*. Quadrangle Books, Chicago.

Likens, G.E., F.H. Bormann, and N.M. Johnson (1969). Nitrification: important to nutrient losses from a cutover forested ecosystem. *Science* **163**:1205–1206.

Odum, E.P. (1969a). The strategy of ecosystem development. *Science* **164**:262–270.

_____. (1969b). Air–land–water — an ecological whole. *J. Soil Water Conserv.*, **24**:4–7.

Essay 10:
Natural Areas as Necessary Components of Man's Total Environment
Eugene P. Odum and Howard T. Odum

Lugo, Ariel E., Snedaker, S.C., Bayley, Suzanne, and Odum, H.T. (1971). *Models for Planning and Research for the South Florida Environmental Study*. Final Report, Contract 14-10-9-900-363, National Park Service/Center for Aquatic Sciences, University of Florida, Gainesville.

Odum, Eugene P. (1970). Optimum population and environment: a Georgian microcosm. *Current History* **58**:355–359.

Odum, Howard T. (1967). Biological circuits and marine systems of Texas. In: *Pollution and Marine Ecology* (T.A. Olson and F.V. Burgess, eds.), pp. 99–151. John Wiley and Sons, New York.

_____. (1968). Work circuits and systems stress. In: *Symposium on Primary Productivity and Mineral Cycling in Natural Ecosystems* (H.E. Young, ed.), pp. 81–138. University of Maine Press, Orono.

_____. (1971). *Environment, Power and Society.* John Wiley and Sons, New York.

Parlzek, R.R., Kardos, L.T., Sopper, W.E., Meyers, E.A., Davis, D.E., Farrell, M.A., and Nesbit, J.D. (1967). *Waste Water Renervation and Conservation.* Pennsylvania State University Studies, University Park.

Sopper, William E. (1968). Waste water renervation for reuse. In: *Water Research,* Vol. 2, pp. 471–480. Pergamon Press, New York.

Essay 11:
Energy, Ecosystem, Development and
Environmental Risk
Eugene P. Odum

Barkley, Paul W., and D.W. Seckler (1972). *Economic Growth and Environmental Decay.* Harcourt, Brace, Jovanovich, New York.

Belt Jr., C.B. (1975). The 1973 flood and man's constriction of the Mississippi River. *Science,* **189**:681–684.

den Boer, P.J. (1968). Spreading of risk and stabilization of numbers. *Acta Biotheor.* **18**:165–194.

Dolan, R., P.J. Godfrey, and W.E. Odum (1973). Man's impact on the barrier islands of North Carolina. *Am. Scient.* **61**:152–162.

Georgescu-Roegan, Nicholas (1971). *The Entropy Law and the Economic Process.* Harvard University Press, Cambridge.

Hafele, Wolf (1974). A systems approach to energy. *Am. Scient.* **62**:438–447.

Meadows, D.H., D.L. Meadows, J. Randers, and W.W. Behrens III (1972). *The Limits of Growth.* Universe Books, New York.

Meadows, P.L., and D.H. Meadows, eds. (1973). *Towards Global Equilibrium: collected papers.* Wright-Allen Press, Cambridge.

Miller, James G. (1965). The organization and life. *Persp. Biol. Med.* **9**: 107–125.

The new math for figuring energy costs. *Business Week,* 8 June 1974, pp. 88–89.

Odum, E.P. (1969). The strategy of ecosystem development. *Science* **164**:262–270.

_____. (1975). *Ecology,* 2nd. ed. Holt, Rinehart & Winston, New York.

Odum, E.P., and H.T. Odum (1972). Natural areas as necessary components of man's total environment. *Trans. 37th N.A. Wildl. Nat. Res. Conf.,* pp. 178–189. Wildlife Management Institute, Washington, DC.

Oertel, George (1973). F. observation on net shoreline position and approximations of barrier island sediment budgets. *Technical Report 78-2.* Georgia Marine Science Center, Skidaway Island.

Reddingius, J., and P. den Boer (1970). Simulation experiments illustrating stabilization of animal numbers by spreading the risk. *Oecologia,* **5**: 240–284.

Ridker, R.G. (1972). Population and pollution in the United States. *Science* **176**:1085–1090.

_____. (1973). To grow or not to grow: that's not the relevant question. *Science,* **182**:1315–1318.

Steinhart, John S., and C.E. Steinhart (1974). Energy use in the U.S. food system. *Science* **184**:307–316.

Essay 12:
The Transition from Youth to Maturity in Nature
and Society
Eugene P. Odum

Axelrod, Robert (1984). *The Evolution of Cooperation.* Basic Books, New York.

Jordan, Carl (1982). Amazonian rain forests. *Am. Scient.* **70**:394–401.

Muscatine, Leonard, and James Porter (1977). Reef corals: mutualistic symbiosis adapted to nutrient-poor environments. *Bioscience* **27**:394–401.

Newell, Sandra, and Elliot Tramer (1965). Reproductive strategies of herbaceous plants in ecological succession. *Ecology* **59**:228–234.

Odum, Eugene P. (1983). *Basic Ecology.* Saunders College Publishing, Philadelphia.

_____. (1989a). *Ecology and Our Endangered Life-Support Systems.* Sinauer Associates, Sunderland, MA.

_____. (1989b). Bridging the four major "gaps" that threaten human and environmental quality. *BRIDGES: An Interdisciplinary Journal of Theology, Philosophy, History, and Science* **1**(3/4):135–141.

Repetto, Robert, et al. (1989). *Wasted Assets.* World Resources Institute, Washington, DC.

Wilde, S.A. (1968). Mycorrhizae and tree nutrition. *Bioscience* **18**:482–484.

Essay 13:
Reduced-Input Agriculture Reduces Nonpoint Pollution
Eugene P. Odum

Adkisson, P.L., C.A. Niles, J.K. Walker, L.S. Bird, and H.B. Scott (1982). Controlling cotton's insect pests, a new system. *Science* 21:19–22.

Buttel, Frederick H., G.W. Gillespie Jr., R. Janke, B. Caldwell, and M. Sarrantonio (1987). Reduced-input agricultural systems: rational prospects. *Am. J. Alternative Agric.* 1(2):58–64.

Frye, W.W., W.G. Smith, and B.J. Williams (1985). Economics of winter cover crops as a source of nitrogen for no-till corn. *J. Soil Water Conserv.* 40(2):246–249.

Gebhardt, M.R., T.C. Daniel, E.E. Schweizer, and R.R. Allmaras (1985). Conservation tillage. *Science* 230:625–630.

Groffman, Peter M., P.F. Hendrix, and D.A. Crossley (1987). Nitrogen dynamics and no-tillage agroecosystems with inorganic fertilizer or legume inputs. *Plant Soil* 97:315–332.

Smith, R.A., R.B. Alexander, and M.G. Wolman (1987). Water quality trends in the nation's rivers. *Science* 235:1607–1615.

Tangley, Laura (1986). Crop productivity revisited. *BioScience* 36:142–147.

White, Fred C., J.R. Hairston, W.N. Musser, H.F. Perkins, and J.F. Reed (1981). Relationship between increased crop acreage and nonpoint pollution: a Georgia case study. *J. Soil Water Conserv.* 36(3):172–177.

Essay 14:
When to Confront and When to Cooperate
Eugene P. Odum

Keddy, Paul (1990). Is mutualism really irrelevant to ecology? *Bull. Ecolog. Soc. Am.* 71(2).

Kropotkin, Peter (1902). *Mutual Aid: A Factor in Evolution.* William Heinemann, London. Reprinted 1935, Horizon Books, Boston.

Essay 17:
Diversity and the Survival of the Ecosystem
Eugene P. Odum

den Boer, P.J. (1968). Spreading of risk and stabilization of animal numbers. *Acta Biotheor.* 18:165–194.

Fiebleman, J.K. (1954). Theory of integrative levels. *Br. J. Phil. Sci.* **5**:59–66.

Grobstein, D. (1969). Organizational levels and explanation. *J. Hist. Biol.* **2**:199–206.

Gutierrez, L.T., and W.R. Fey (1975). Feedback dynamics analysis of secondary successional transients in ecosystems. *Proc. Natl. Acad. Sci. USA* **72**:2733–2737.

Odum, E.P. (1975a). Diversity as a function of energy flow. In *Unifying Concepts in Ecology. Proc. First Int. Congr. Ecol.* (W.H. van Dobben and R.H. Lowe-McConnell, eds.), pp. 11–14. W. Junke Publishers, The Hague.

Odum, E.P. (1975b). *Ecology, the Link Between the Natural and the Social Sciences.* Holt, Rinehart & Winston, New York.

Reddingius, J., and P.J. den Boer (1970). Simulation experiments illustrating stabilization of animal numbers by spreading the risk. *Oecologia* **5**:240–284.

Redfield, R., ed. (1942). *Levels of Integration in Biological and Social Systems.* Biological Symposia, No. 8. Cattell Press, Lancaster, PA.

Essay 19:
Energy, Ecology, and Economics
Howard T. Odum

Lotka, A.J. (1922). Contribution to the energetics of evolution. *Proc. Natl. Acad. Sci. USA* **8**:147–188.

Meadows, D.H., D.L. Meadows, J. Randers, and W.W. Behrens III (1972). *The Limits to Growth.* Universe Books, New York.

Odum, H.T. (1971). *Environment, Power and Society.* John Wiley, New York.

Essay 21:
Environmental Degradation and the Tyranny of Small Decisions
William E. Odum

Kahn, Alfred E. (1966). The tyranny of small decisions: market failures, imperfection and the limits of economics. *Kyklos* **19**:47.

Essay 22:
**Bridging the Four Major "Gaps" that Threaten Human
and Environmental Quality**
Eugene P. Odum

Gebhardt, M.R., T.C. Daniel, E.E. Schweizer, and R.R. Allmaras (1985). Conservation tillage. *Science* **230**:625–630.

Hawkins, Paul, James Ogilvy, and Peter Schwartz (1982). *Seven Tomorrows: Towards a Voluntary History*. Bantam, New York.

Kahn, Alfred E. (1966). The tyranny of small decisions: market failures, imperfection, and the limit of economics. *Kyklos* 19:23–47.

Little, C.E. (1987). *Green Fields Forever: The Conservation Tillage Revolution in America*. Island Press, Washington, DC.

Meadows, D.H., D.L. Meadows, J. Randers, and W.W. Behrens III (1972). *The Limits of Growth*. Universe Books, New York.

Odum, Eugene P. (1989a). *Ecology and Our Endangered Life-Support Systems*. Sinauer Associates, Sunderland, MA.

_____. (1989b). Input management of production systems. *Science* **243**:177–182.

Odum, William E. (1982). Environmental degradation and the tyranny of small decisions. *BioScience* **42**:412–414.

Repetto, Robert, et al. (1989). *Wasted Assets*. World Resources Institute, Washington, DC.

Seligson, Mitchell A. (1984). *The Gap Between the Rich and the Poor*. Westview Press, Boulder, CO.

Spencer, D.F., S.B. Alpert, and H.H. Gilman (1986). Cool water: demonstration of a clean and efficient new coal technology. *Science* **232**:609–612.

UNFAO (1985). *FAO Production Yearbook*, Vol. 39. United Nations Food and Agricultural Organization, Paris.

Essay 24:
**Source Reduction, Input Management, and
Dual Capitalism**
Eugene P. Odum

Gunn, J.M., ed. (1995). *Restoration and Recovery of an Industrial Region: Progress in Restoring the Smelter-Damaged Landscape Near Sudbury, Canada*. Springer-Verlag, NY.

Kirsch, F.W., G.P. Looby, and M.E. Kirk. (1993). Case studies: how four manufacturers improved painting operation to reduce waste. *Pollution Prevention Review*, Autumn, pp. 429–436.

Odum, E.P. (1989). Input management of production systems. *Science* **143**:177–182.

Index

A

Acacia trees and ants, 35

Acid rain, 48, 49

Acreage requirements, 75–76, 139–141

Affluence and supported population size, 8, 74

Agriculture
and biodiversity, 40
demands on, 75
drainage, 127–128
industrialized, 48
as mutualistic relationship, 35
output, 169–172
shifting, 98
as solar-powered ecosystem, 108–109
traditional, 40

Air pollution, gradual, 223

Algae and coral reefs in mutualistic relationship, 95–96, 173

Ants
and acacia trees, 35
and population control, 4–5

Aquaculture, 108–109, 196

Atomic energy, use of, 232

B

Balance of nature, 27–28

Balance of payments of a country, 211–212

Beach as an ecosystem, 159–161

Beach erosion, 52 (Cartoon 7.4)

Behavior and ecology, 33–37

Between habitat diversity, 183

Biodiversity, 39–40, 147–148, 187–191, 246–247
preservation of, 181–185

Biological polymorphism, 83

Biosphere services, 56 (Fig. 8.1)

Biotechnology research in the university, 232–234

Biotic communities, development of, 164–166

Birds and confrontation, 174–175

Birth control, 77

Bond energies, 122

Boom-and-bust pattern of growth, 2–4, 55, 101, 147, 218

Boulding, Kenneth, 62

Breeder process in nuclear energy, 210–211

Butzer, Karl, 21, 178

Dual capitalism, 45–46, 55, 236
Duncan's "Ecological Complex,"
    80–81

E

Earth stewardship, 237–240
Easter Island, 37
Ecological hierarchy, 23–25
    (Table 3.1)
Ecological succession, 19, 20
    (Fig. 2.3), 164–165, 188, 189
Ecology
    and behavior, 33–37
    and change, 27
    and diversity, 39–40, 147–148,
        187–191, 246–247
    and economics, 43–46, 125, 218
    and energy, 13–21, 149, 199–215
    and growth, 1–12
    and humans, 80–82
    and organization, 23–26
    and sociology, 79–80
    to solve problems, 241–248
    views of, 80–82
    and the watershed, 127–131
Economic development, 102
Economic externalities, 45
Economic growth, 62–63
Economics
    and ecology, 125, 146, 218
    of scale, 67
Ecosystem, 23–25
    concept of, 123–125, 149–150
    control mechanism, 121–122
    development of, 29–30, 62
    and diversity, 187–191
    and energy, 157–158
    external dynamics, 119–121
    management of, 25, 115–125,
        136–138

and population, 73–75
power requirements of, 69
Education gap, 228–229
Electricity, production of, 15
Emergent property principle, 26,
    117–119. *See also* Holism
EMergy, 45
Emissions, 49
Energetics, 13, 15
Energy
    auxiliary, 205–206
    concentration of, 15, 17–18,
        208–210
    conservation, 245
    consumption in industrial areas,
        109–110
    conversion, 65, 154–155, 244–245
    cost of, 153–156, 180
    dependence upon, 13–14
    disorder, 14, 66, 155
    dissipation, 106
    and ecology, 149, 199–215
    efficiency, 204, 207
    embodied, 45
    growth, 165
    laws, 14–17, 106
    maintenance, 165–166
    measurement of, 105–106
    not reusable 13, 16 (Cartoon 2.1)
    quality of, 17–18, 106, 210
    reserves, calculation of, 203
    and society, 65–71
    and steady state growth, 204–205
    storage, 211
    subsidies, 205–211
    systems, 107–111
    transformation 13–14 (Fig. 2.2),
        106
        and concentration, 17–18
    units of, 105–106
Energy density of the land, 123–125